Michael Oldenburg

Photon upconversion heterostructures made from surface-anchored metal-organic frameworks

Schriften des Instituts für Mikrostrukturtechnik
am Karlsruher Institut für Technologie (KIT)
Band 40

Hrsg. Institut für Mikrostrukturtechnik

Eine Übersicht aller bisher in dieser Schriftenreihe
erschienenen Bände finden Sie am Ende des Buchs.

Photon upconversion heterostructures made from surface-anchored metal-organic frameworks

by
Michael Oldenburg

KIT Scientific Publishing

Karlsruher Institut für Technologie
Institut für Mikrostrukturtechnik

Photon upconversion heterostructures made from surface-anchored metal-organic frameworks

Zur Erlangung des akademischen Grades eines Doktor-Ingenieurs von der KIT-Fakultät für Elektrotechnik und Informationstechnik des Karlsruher Instituts für Technologie (KIT) genehmigte Dissertation

von Michael Oldenburg

Tag der mündlichen Prüfung: 05. März 2018
Erster Gutachter: Prof. Dr. Bryce S. Richards
Zweiter Gutachter: Prof. Dr. Heinz Kalt

Impressum

KIT Scientific Publishing

Karlsruher Institut für Technologie (KIT)
KIT Scientific Publishing
Straße am Forum 2
D-76131 Karlsruhe

KIT Scientific Publishing is a registered trademark of Karlsruhe Institute of Technology.
Reprint using the book cover is not allowed.

www.ksp.kit.edu

Print on Demand 2019 – Gedruckt auf FSC-zertifiziertem Papier

ISSN 1869-5183
ISBN 978-3-7315-0863-2
DOI 10.5445/KSP/1000087374

Kurzfassung

Materialien, welche absorbiertes Licht in höherenergetisches Licht umwandeln, d.h. welche Photonen-Hochkonversion durchführen, sind vom großen Interesse auf den Gebieten der Energieumwand-lung, der hintergrundfreien Mikroskopie/Bildgebung und der lumineszierenden Markierung. Insbesondere Hochkonversion auf der Basis der Triplett-Triplett-Annihilation (TTA-UC) hat in den letzten Jahren großes Interesse geweckt. Hierbei wird zuerst ein Absorbermolekül mit einer Beleuchtungsquelle mit geeigneter Wellenlänge angeregt. Nach der Anre-gung überträgt das Absorbermolekül seine Energie auf ein Emittermolekül, welches anschließend mit einem anderen Emittermolekül kollidiert und so eine hochenergetische Anregung auf einem der Emittermoleküle erzeugt wird. Der Zerfall des so angeregten Zustands kann zu einer Emission eines Photons führen, welches eine höhere Energie besitzt als die der zuvor absorbierten Photonen. Einer der Gründe für die jüngste Popularität von TTA-UC ist, dass die maximale Effizienz unter sonnenähnlichen Strahlungsin-tensitäten erreicht werden kann. Für technologische Anwendungen ist das eine wichtige Voraussetzung.

Für TTA-UC gibt es zwei Leistungsindikator. Der erste ist die maximale Hochkonversion-Quantenausbeute. Da der Hochkonversionsprozess nichtlinear ist hängt die Hochkonversion-Quantenausbeute bei niedrigen Strahlungsintensitäten linear von der Beleuchtungsintensität ab. Bei ausreichend hohen Strahlungsintensitäten erreicht die Hochkonversion-Quantenausbeute jedoch einen Sättigungswert – die maximale Hochkonversionquantenausbeute. Der zweite Leistungsindikator ist die Strahlungsintensitätsschwelle der Hochkonversion. Die Strahlungsintensitätsschwelle kennzeichnet eine charakteristische Strahlungsintensität, welche den Übergang vom nichtlinearen zu einem linearen Prozess markiert.

Ein gutes TTA-UC System hat eine hohe Quantenausbeute (maximal jedoch 50%) und eine niedrige Strahlungsintensitätsschwelle. Diese Arbeit analysiert

Materialen in denen TTA-UC stattfindet für zukünftige Anwendung in optischen Bauteilen.

Die theoretischen Grundlagen werden mit Hilfe von Ratengleichungen beschrieben. Diese gelten für eine isotrope Verteilung von Absorber- und Emittermolekülen, wie es in einer Lösung der Fall ist. Zuerst wird das stationäre Verhalten unter konstanter Strahlungsintensität analysiert und intensitätsabhängige Funktionen für beide Leistungsindikatoren hergeleitet. Ferner wird eine detaillierte Analyse für den Fall einer Rechteckwellenanregung durchgeführt. Das so gefundene theoretische Verhalten wird anhand eines Standardsystems verifiziert. Basierend auf der Bestätigung der Theorie durch experimentelle Daten wird eine neue Hypothese für eine nicht isotrope Verteilung von Absorber und Emittermolekül aufgestellt. Es wird ein Doppel-schichtsystem angenommen, bei dem eine Schicht ausschließlich aus dem Absorbermolekül und die zweite Schicht aus dem Emittermolekül besteht. Innerhalb dieses Doppelschichtsystems hängt die Strahlungsintensitätsschwelle von der Dicke der Absorber- und der Emitterschicht ab. Die Schlussfolgerung ist, dass die Absorberschicht so dick wie möglich, aber die Emitterschicht so dünn wie möglich sein muss. Dadurch werden angeregte Zustände von der Absorberschicht in der Emitterschicht konzentriert und somit wird die Strahlungsintensitätsschwelle gesenkt.

Eine neue Klasse von synthetischem Material wird verwendet, um dieses Verhalten zu überprüfen. Oberflächenverankerte metallorganische Gerüstmaterialien (SURMOF) sind molekulare Systeme, in welchen Metallatome organische Linker zu einer kristallinen Struktur koordinieren. SURMOFs lassen sich sehr einfach als Dünnschichtfilme auftragen und eignen sich daher ideal für die Anwendung in Bauteilen. Durch eine zyklusweise Auftragung der Metall- und der Linkerlösung auf ein flaches Substrat können große Flächen von organischen Molekülen, die in einer geordneten Weise ausgerichtet sind, hergestellt werden. Die kristalline Struktur begünstigt die Transporteigenschaften von angeregten Zuständen.

Ein Emitter SURMOF (ADB SURMOF) wird basierend auf dem organischen Molekül Diphenylanthracen hergestellt, welches zu den besten Emittermolekülen für TTA-UC gehört. Die spektroskopische Charakterisierung führt zu einem besseren Verständnis der Energietransporteigenschaften innerhalb des Materials. Zeitaufgelöste Photolumineszenz-Daten zeigen zwei angeregte Zustände innerhalb des Emittersystems. Einer dieser Zustände ist für eine niedrige Photolumineszenzquantenausbeute (PLQY) dieses Systems verantwortlich, is den Emitter SURMOF nicht ideal für TTA-UC macht. Durch Zugabe des Farbstoffs Cumarin 343 in das Gerüst kann die PLQY jedoch von unter 5% auf bis zu 52% gesteigert werden. Weiterhin konnte eine Diffusionslänge der angeregten Zustände von ungefähr 60 nm abgeschätzt werden.

Die Absorber-SURMOF-Schicht basierend auf einem Pd(II)-Porphyrinmolekül wurde analysiert, das Pd-DCP SURMOF. Die Haupteigenschaften, welche relevant für TTA-UC sind wurden gemessen: erstens, die Intersystem-Crossing-Effizienz, welche die Effizienz des Übergangs von direkt angeregten Singulettzuständen in langlebige Triplettzustände beschreibt und zweitens, die Lebensdauer der so gebildeten Triplettzustände. Die Intersystem-Crossing-Effizienz liegt für den Absorber bei fast 100% und die Lebensdauer der Triplettzustände beträgt ca. 300 µs. Beide Parameter machen den Absorber SURMOF zu einem idealen Kandidaten für TTA-UC.

Zwei Arten von Schichtstrukturen für TTA-UC wurden auf der Basis von SURMOFs hergestellt. Ein System bestehend aus drei Schichten, in welchem der ADB SURMOF zwischen zwei Pd-DCP SURMOFs liegt. In diesem System wurde die Emission von blauem Licht bei ca. 480 nm nach einer Anregung bei ca. 532 nm beobachtet, welches bestätigt, dass TTA-UC in diesem System funktioniert. Weiterhin dient dies auch als Bestätigung für die Übertragung von Elektronen an der Grenzschicht zwischen ADB und Pd-DCP SURMOF. Jedoch konnte auch beobachtet werden, dass der Intensitätsschwellenwert zusammen mit der Dicke der Absorberschicht anstieg. Dieses Problem konnte in Verbindung mit einer schlechten Grenzfläche zwischen Absorber- und Emitter-SURMOF in Verbindung gebracht werden. Mittels

Ultraschalles konnte die Grenzfläche verbessert und das theoretisch vorhergesagte Verhalten für den Schwellenwert verifiziert werden. Intensitätsschwellenwerte von weniger als 1 mW cm^{-2} konnten gemessen werden. Obwohl die Schwelle extrem niedrig war, betrug die Upconversion-Quantenausbeute nur ca. 0.5%.

Im nächsten Optimierungsschritt wurde der Farbstoff Cumarin 343 innerhalb des Doppelschichtsystems dem ADB SURMOF zugefügt, um die Hochkonverions-Quantenausbeute auf bis zu 5% zu erhöhen. Es wurde jedoch beobachtet, dass aufgrund der sehr guten Transporteigenschaften der Anregungen innerhalb des SURMOFs die meisten hochkonvertierten Anregungen wieder zurück an den Pd-DCP SURMOF abgegeben werden. Die erarbeiteten Ergeb-nisse und Analysen zeigen die derzeitigen Einschränkungen für den Einsatz von SURMOFs in optischen Bauteilen auf, welche mit der molekularen Struktur zusammenhängen. Für zukünftige Forschungsarbeiten werden diese Erkenntnisse eine entscheidende Rolle spielen.

Abstract

Material systems that perform photon upconversion (UC), emitting photons of higher energy than those with which they are excited, are of interest in the fields of energy conversion, background-free microscopy/imaging, and luminescent marking. Especially, upconversion based on triplet-triplet annihilation (TTA-UC) gained significant attention in the last decade. In this process, the absorber molecule gets excited with an illumination source of an appropriate wavelength. After excitation, an emitter molecule accepts the energy of the absorber molecule. When two excited emitter molecules collide, their energy combines, and a high-energy excitation is created. The relaxation of this new excited state leads to the emission of a high-energy photon. One of the reasons for the recent popularity of TTA-UC is that it reaches its maximum efficiency under sunlight illumination, which is preferable for many applications.

There are two figures-of-merit for TTA-UC. First, the maximum upconversion quantum yield (UCQY). The UCQY depends linearly on the illumination intensity for low excitation power densities. However, at sufficiently high excitation power densities the UCQY reaches a saturation value referred to as the maximum UCQY. The second figure-of-merit is the UC threshold. The UC threshold marks a fixed excitation power density that signifies the transition from a non-linear to a linear process. A good TTA-UC system has a high UCQY (max. 50%) and a low UC threshold. In this thesis, TTA-UC materials are investigated that fulfill both requirements for potential future optical devices.

The theoretical foundations are described using rate equations that are based on an isotropic distribution of absorber and emitter molecules, like in a solution. First, the steady-state behavior under constant illumination is analyzed, and intensity dependent functions for both, the maximum UCQY and the UC threshold are derived. Further, a detailed analysis is carried out for the case of a square wave excitation.

Both theoretical approaches are verified for a standard UC system in solution. Based on the verification of the theory by experimental data a new hypothesis is stated for a non-isotropic distribution of absorber and emitter molecule. A bilayer system is assumed in which one layer is made purely of the absorber molecule and the second layer of the emitter molecule. Within this bilayer system, the UC threshold depends on the thickness of the absorber and the emitter layer. The conclusion is that the absorber layer must be as thick as possible, while the emitter layer should remain as thin as possible. Via this approach, excited states from the absorber layer are concentrated within the emitter layer and thus decrease the threshold of the system.

A new class of synthetic material is used to verify this behavior. Surface-anchored metal-organic frame-works (SURMOF) are molecular systems in which metal centers are coordinating organic linkers into a crystalline structure. SURMOFs are easily grown as thin-films making them ideal for the application in devices. By cycle-wise deposition of the metal and the linker solution onto a flat substrate, large areas of organic molecules that are aligned in an ordered manner can be grown. The crystalline structure favors the transport properties of excited states.

An emitter SURMOF is build based on the organic molecule diphenylanthracene that is one of the best performing emitters for TTA-UC. Spectroscopic characterization leads to a better understanding of the energy transport properties of the material. From time-resolved data, two excited states are found. One of these states is responsible for a low photoluminescent quantum yield (PLQY) of this system, making the emitter SURMOF not ideal for TTA-UC. However, by adding the guest molecule coumarin 343 into the framework, the PLQY could be increased from below 5% up to 52%. Along with this substantial PLQY enhancement, the diffusion length of excited states is estimated to be around 60 nm long.

The absorber SURMOF layer based on a Pd(II) porphyrin molecule is analyzed (Pd-DCP SURMOF). The main properties for TTA-UC are measured: i) the ISC efficiency, that states the efficiency of the transition from directly excited singlet states into long-living triplet states and ii) the lifetime of the so creat-ed triplet states. The ISC efficiency is found to be close to 100% and the lifetime of the triplet states is found to be ~300 µs. Both parameters making the absorber SURMOF an ideal candidate for TTA-UC.

Optical devices based on two types of layered structures are made. A three layer in which the emitter is sandwiched between two absorber layers. With this strategy, the excited states of both absorber layers are concentrated in the emitter layer. In this three-layer, TTA-UC could be verified by the emission of blue light around ~480 nm after excitation at 532 nm. The observation of TTA-UC also serves as a proof that electrons cross the interface between ADB and Pd-DCP SURMOF. However, it is found that the threshold increases together with the absorber layer due to bad triplet transfer across the heterostructure. This issue is correlated to a bad interface across the heterostructure. By ultrasonication, the interface could be improved, and the predicted behavior for the threshold is verified. Thresholds of <1 mW cm^{-2} are found, making SURMOFs ideal for application.

However, the UCQY is less than 0.5% in the analyzed bilayers. In the next step of optimization, the coumarin 343 dye is added to the emitter layer. A simple calculation shows that this should increase the UCQY up to 5%. Yet, it is found that due to the excellent transport properties of the SURMOFs high energy excited states that are generated are back transferred to the absorber layer and cannot increase the UCQY. The obtained results and analyzes show the current limitations for the use of SURMOFs in optical devices, which are related to the molecular structure. For future research, these findings will play a crucial role.

Acknowledgements

Many people and institutions contributed to the success of this work and SURMOFS!

Big thanks go to Natalia Kiseleva my significant other, who I met during my PhD and who is one big reason why this PhD was actually a very good idea. I appreciate your patience and your very positive attitude that helped me tremendously during my PhD.

I would like to thank Ian Howard. I never met a person with such an enormous amount of patience. I think it is not the easiest to supervise someone who is sometimes very emotionally driven. Thank you for guiding me through this work and keeping me on track. I really hope that our work and effort will have a strong impact on the research of organic electronics!

Also, I would like to thank my Doktorvater Bryce Richards who gave me the chance for a PhD in the field of experimental science although coming from theoretical physics. The amount of trust he places into people is enormous and I really hope that everyone will appreciate this as much as I do. Thank you also for organizing oral talks in every conference I have been to! I hope that I have communicated our work and ideas well to the rest of the world.

Many thanks are going to my colleagues, Stephan Dottermusch, Deski Beri, Michael Adams, Sheying Li, Dmitry Busko and Andrey Turshatov for having a fun time in the institute and for discussing any ideas I had during every single day of my PhD. Special thanks go to Nicolo Baroni for teaching me the Italian culture, going with me on Halloween as Mario (!) and for sharing with me the problems that arose from SURMOFs!

I would like to thank the very competent people from the IFG laboratories: Peter Weidler, Stefan Heisler, Alexander Well, Ritesh Haldar, Christof Wöll and Engelbert Redel. Thank you for instructive critics and for the access to the laboratories.

I would like to thank my family for supporting me, although not always knowing what I am actually doing.

The last thank you goes to the people from the Victoria University in Wellington, New Zealand, Justin Hodgkiss and Kai Chen. It was a fun time with you. Other thanks go to Adrian and Earl who accommodated me in their very cozy house and for worrying so much that I might get bored in New Zealand. The research stay abroad was funded by the Karlsruhe House of Young Scientists (KHYS).

List of journal publications

M. Oldenburg, A. Turshatov, D. Busko, S. Wollgarten, M. Adams, N. Baroni, A. Welle, E. Redel, C. Wöll, B. S. Richards and I.A. Howard, Photon Upconversion at Crystalline Organic–Organic Heterojunctions. Adv. Ma-ter., 2016, 28: 8477–8482.

N. Baroni, A. Turshatov, M. Oldenburg, D. Busko, M. Adams, R. Haldar, A. Welle, E. Redel, C. Wöll, B. S. Richards, I. A. Howard, Facile loading of thin-film surface-anchored metal-organic frameworks with Lewis-base guest molecules. Mater. Chem. Front., 2017,1, 1888-1894.

N. Kiseleva, M. A. Filatov, M. Oldenburg, D. Busko, M. Jakoby, I. A. Howard, B. S. Richards, M. Senge, S. M. Borisov, A. Turshatov, The Janus-Faced Chromophore: A Donor-Acceptor Dyad with Dual Performance in Photon Up-conversion. Chem. Commun., 2018, Accepted Manuscript.

M. Oldenburg, A. Turshatov, D. Busko, M. Jakoby, R. Haldar, K. Chen, G. Emandi, M. O. Senge, C. Wöll, J. M. Hodgkiss, B. S. Richards and I. A. Howard, Enhancing the photoluminescence of surface anchored metal–organic frameworks: mixed linkers and efficient acceptors, Phys. Chem. Chem. Phys., 2018, DOI: 10.1039/c7cp08452h.

M. Oldenburg, R. E. Joseph, N. Kiseleva, D. Kim, D. Busko, B. S. Richards, I. A. Howard, Triplet-Triplet Annihilation Upconversion Threshold Determination: A Comparison of Steady-State and Time-Resolved Methods, in preparation.

List of conference contributions

M. Oldenburg, A. Turshatov, D. Busko, S. Wollgarten, M. Adams, N. Baroni, A. Welle, E. Redel, C. Wöll, B. S. Richards and I.A. Howard, Elec-tronic transfer across crystalline thin-film surface anchored metal-organic framework (SURMOF) heterojunctions. SPIE Photonics Europe, 3 - 7 April 2016 in Brussels, Belgium

M. Oldenburg, A. Turshatov, D. Busko, C. Wöll, B. S. Richards and I. A. Howard, Reducing triplet-triplet annihilation upconversion thresholds in thin films via spatial confinement of excitons in organized chromophore layers. AMN8, 12 - 16 February 2017 in Queenstown, New Zealand

M. Oldenburg, D. Busko, R. Joseph, D. Kim, D. Hudry, B. S. Richards, I. A. Howard, Up-conversion Thresholds: Describing the Power Dependence of Up-Conversion from Triplet Annihilation Systems to Lanthanide Core-Shell Nanoparticles. SHIFT 2017, 13 - 17 November 2017 in Costa Adeje, Tenerife, Spain

Table of contents

1 Background

1.1 Motivation

The sun has long been central to human civilization. In ancient cultures, the sun is assumed to be the materialization of a god and solar eclipses are correlated to human misbehavior. Although our belief system has changed over the years, the fascination for the sun remained. Now, in times of climate change and the Paris Agreement this fascination transformed into a strong ecological reliance. Solar energy harvesting technologies offer the possibility of emission-free and sustainable energy production. Solar cells, which can convert light into electrical energy, are currently one bearer of hope for many nations that aim to reduce greenhouse gas emission from coal power plants.

The basic principle of a solar cell is that electrons in a semiconductor are excited from states in the valence band to states in the conduction band via the absorption of light. The energetic difference between the valence and conduction bands is the bandgap. Electrons can only get excited into the conduction band by photons that have at least the energy of the bandgap. The excitation of an electron from the valence band to the conduction band also leaves an empty state, a hole, in the valence band. Conduction band electronics and valence band holes can be then extracted from the semi-conductor do work in an external circuit. Silicon-based solar cells lead the commercial market. The production share of silicon solar cells is 93% in 2016 and the installed system prices dropped from 14000€/kWp in 1990 to 1270€/kWp in 2016.[1] The immense price drop is mostly due to effect of

[1] http://www. ise.fraunhofer.de/content/dam/ise/de/documents/publications/studies/Photovoltaics-Report.pdf

scale, which means that the number of production sites increased and the production itself is highly optimized. This makes silicon difficult to be replaced by another material in the future.

However, a single bandgap material like silicon with a bandgap of 1.12eV suffers from several loss mechanisms. The two main loss mechanisms are thermalization and transmission losses. The origin of thermalization losses are photons that exceed the bandgap energy. Absorbed light that exceeds the energy of the bandgap creates "hot" carriers that quickly relax to the conduction band. The difference in energy between the absorbed photon and the conduction and valence band energy goes into lattice vibrations/phonons. This loss is referred to as thermalization loss i.e heat. Transmission losses are due to photons that have lower energy than the bandgap. A single bandgap material is transparent for these photons and so no electrical energy can be generated. Figure 1.1 shows the contribution of both transmission and thermalization losses as a function of the bandgap of the absorber.

One strategy to overcome the thermalization and transmission losses is to use several absorbing materials with different bandgap whereby the material with the highest bandgap is on top. With this approach the thermalization losses are kept small, as high energy photons are absorbed by a high bandgap material and transmission losses are reduced as low energy photons are absorbed by a low bandgap material. The tandem cell approach has been demonstrated to reach very high conversion efficiencies. Conversion efficiencies of up to 41.1% have been demonstrated by Guter et al by using triple-junction solar cell based on GaInP/GaInAs/Ge under a irradiation of 454 suns. [2] Additionally, Essig et al demonstrated a mechanically stacked three junctions module based on a dual junction cell GaInP/GaAs and a Si single junction cell with a conversion efficiency of 35.9%. [3] These two example demonstrate one strong loss mechanism in tandem cells, the lattice mismatch between Si and the III-V semiconductors that leads to charge carrier trapping and so reduce the conversion efficiency. A promising partner material that can be used together with Si is perovskite. Recent work was done by Bush et al that created a monolithic perovskite/Si tandem solar cell that has a conversion efficiency of 23.6%. [4] Although this is lower than the current record conversion efficiency of a single-junction Si solar cell (26%) demonstrated by Yoshikawa et al. [5] theoretically a perovskite/Si tandem cell can reach efficiencies of ~40%. [6]

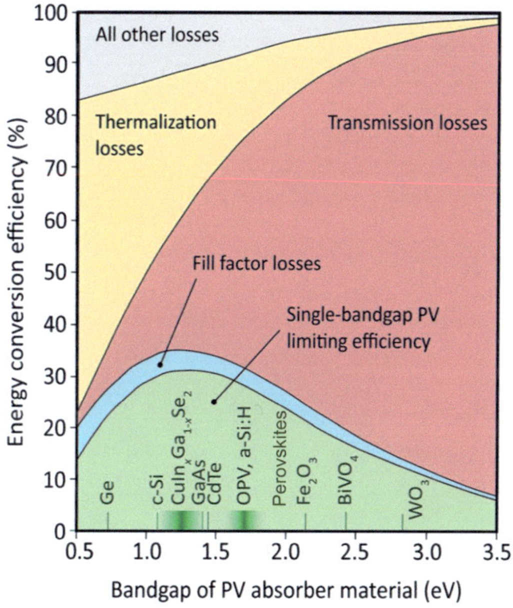

Figure 1.1: Relative impact of fundamental loss mechanisms in solar energy conversion as a function of the PV absorber bandgap, and examples of absorber materials. Copied with permission from Schulze and Schmidt. [1]

Interestingly, very similar loss mechanisms occur for another energy generation process which uses light as an activation step. Photocatalysis which describes the activation of a chemical reaction by light. A very prominent example for a photocatalytic process is the water-splitting into H_2 and O_2. Luckily this process happens extremely slow by pure light illumination and needs an electronic potential difference induced by an excited electron. For this process TiO_2 that has a bandgap of 3.2eV (~380nm) must be excited by ultraviolet (UV) light. [7] Water molecules that are close to the surface of TiO_2 are then split into H_2 and O_2. However, the amount of light with a wavelength lower than 400 nm is only 10%. Figure 1.2 shows a scheme of a possible photocatalytic water splitting device. The so produced hydrogen can then either be used directly in a fuel cell [8] for generation of electricity or as a reagent to make together with carbon dioxide gases like methane. [9] However, as in solar cells the amount of sunlight that is used is limited by the bandgap of the absorber material, here TiO_2.

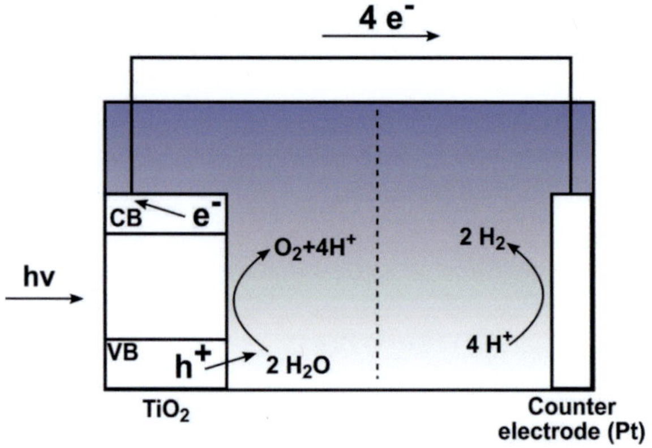

Figure 1.2: A schematic of a photocatalytic water splitting reaction. A photon with an energy higher than the bandgap of TiO_2 excites electrons into the conduction band (CB), the remaining holes in the valence band (VB) are reacting with water molecules and produce oxygen and protons. The protons float to the counter electrode (made from Pt) passing a semipermeable membrane (dashed line) and are reacting with the electrons to form hydrogen.

In the case of solar cells this problem could be solved by using the tandem approach. In the case of water splitting this cannot be done in the same fashion as an electron has to be excited in any case to a high enough potential.

A strategy to overcome the above-mentioned losses is by using spectral conversion materials. For thermalization losses downconversion or quantum cutting is a promising approach. The main idea is that photons that exceed the bandgap energy are absorbed by a downconverter material. Within this material the energy of the absorbed photons is split into two lower energy photons. Ideally the downconverter material is attached on top of the active solar cell material. The splitting of the initial photon into two lower energetic photons allows the generation of two electrons instead of one electron that loses its energy via thermalization.

The transmission losses can be adressed by an opposite strategy, that of UC. Photons with less energy than the bandgap are absorbed by an upconverting material. Within the upconverter the energy of two photons is combined to yield one higher energy photon. The so-created photons can now be absorbed by the active solar cell material and increase the energy conversion efficiency for the spectral region that has lower energy than the bandgap. The UC process relies heavily on the absorption of at least two photons, which makes it a non-linear process. Yet, at high illumination intensities, the process saturates and becomes linear. The aim of this work is to design an UC material for which this transition occurs at sun light illumination that makes it viable for applications.

Currently there are mainly two types of material systems for upconversion. First, upconversion based on inorganic materials. Second, upconversion based on organic molecules. Much work has been done in the past to establish upconverters based on inorganic materials, e.g. rare earth ions. Bloembergen has proposed upconversion in 1958, [10] and it is experimentally verified by Auzel in 1966 for Ytterbium ions. [11] Currently the highest efficient materials based on rare earth ions are Erbium doped fluorides with an efficiency of 15% for a conversion from 1500 nm to 980 nm. [12]

The second material system for UC is based on organic semiconductors. One type of organic semiconductor, referred to as the "sensitizer", absorbs the excitation light and create excitons. These excitons are transferred to the second material, the "emitter". Here, two excitons can combine by pairwise annihilation, which brings an emitter molecule to a higher excited state. The subsequent radiative decay leads to the emission of a photon that has a higher energy than each of the originally absorbed photons. This type of UC is referred to as triplet-triplet annihilation based upconversion (TTA-UC). Due to the high absorption cross-section of organic molecules efficient upconversion is already possible at sunlight intensities. The currently most efficient organic UC system is reported by Hoseinkhani et al. [13] This system reaches UCQY that are close to the thermodynamic limit efficiency of 50% under an excitation intensity of ~2 mW cm-2. [13] In comparison to this, inorganic UC efficiencies of 15% are reached at ~4 kW cm-2. [12]

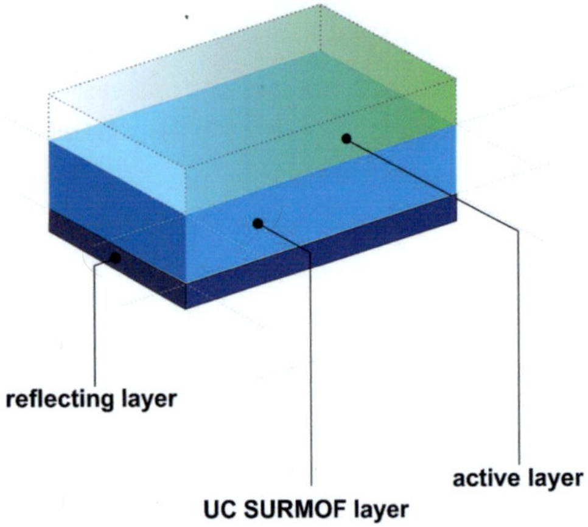

Figure 1.3: A schematic of an active layer that absorbs sun light and is enhanced by an UC layer based on a SURMOF. The reflecting layer at the bottom is reflecting the upconverted light back to the active layer.

This makes organic upconversion systems very promising for applications in the field of solar energy harvesting. However, currently most of the organic upconverting materials are based on a random distribution of sensitizer and emitter molecules in solutions or polymers. This leads to poor energy transport, which lowers the UC efficiency as the collision of two emitter molecules is diffusion limited and so is the process of triplet-triplet annihilation. Solid-state thin films, which are purely made from rationally aligned either sensitizer or emitter molecules, would allow good energy transfer. Furthermore, the concentration of absorbed energy from a thick sensitizer thin film into a very thin emitter film would make upconversion efficient under low intensities. The goal of this work is to investigate an UC system based on surface anchored metal-organic frameworks (SURMOFs). SURMOFs offer an interesting platform for organic upconversion as molecules can be arranged in a crystalline structure. The ordered alignment allows for increased control of the energy transport properties. Furthermore, SURMOFs are inherently thin film structures that can be easily integrated into a device as illustrated very schematically in Figure 1.3. Although the used absorber and emitter molecules that will be used are only useful for water splitting, the studied structures and effects might not differ strongly for other absorber and emitter molecules that are arranged to form SURMOFs. This means that although the current choice of molecules is limited to what currently exists, this can change in the near future.

1.2 Excitons

When matter absorbs light, bound electrons in the material can transition from a low into a high energy level. In the case of organic molecules, in the most simplistic theoretical description the optical gap is determined by the difference between the highest occupied molecular orbit (HOMO) and the first energy level to which the electron can be excited, the lowest unoccupied molecular orbit (LUMO). After the electron is excited into the LUMO level, a vacancy is left in the HOMO level. Consequently, the net charge of this vacancy is positive and so the excited electron is attracted to this "hole". The electron-hole pair forms a bound two-particle system that can be formalized as a single quasiparticle with zero net charge - an exciton. This process depends on the permittivity ϵ_R that is a measure on how easily an electric field can penetrate a material. In inorganic semiconductors ϵ_R is high and the created electron and hole are shielded by the opposite charges strongly. In effect, the attraction between electron and hole is weak and the exciton is extended over many sites within the material. This type of exciton is called Wannier-Mott exciton. For Wannier-Mott excitons, thermal activation is sufficient to dissociate them and create free charges. In contrast to this, in organic semiconductors ϵ_R is small and the electron and the hole charge of electron and hole is only weakly changed. Therefore, the attraction between electron and hole is strong and leads to a strong localization of the exciton. This type of exciton is called Frenkel exciton. As this work is mostly concerned with organic materials, we will consider only Frenkel excitons throughout the remainder of this work. The creation of an exciton by the absorption of one photon is illustrated in Figure 1.4.

Electrons are fermions and possess a negative charge and a spin of ½. Electromagnetic fields can therefore interact via the dipole moment and the magnetic moment with the electron. That would imply that the absorption of a photon can result in a change of the electronic and the spin component of the wavefunction. As the wavefunction of fermions is per definition asymmetric the simultaneous change of the electronic and spin state is forbidden. Therefore, the absorption of a photon cannot change the spin state of the electron. As a result, the exciton is initially always in a singlet state that is labeled by S_1 in Figure 1.4 (given the ground state has a singlet character, which is the case for all materials considered within this thesis). However, as the exciton is not isolated from its environment, perturbations can allow a spin flip after the excitation. This "crossing over" of a singlet to a triplet state is called "intersystem crossing" (ISC) and is first described by Kasha. [14] Furthermore, McClure has shown that the probability of a triplet-to-singlet or singlet-to-triplet transition increases when a heavy atom is added to the molecule an effect that we capitalize upon in this work. [15] ISC is generally explained by the magnetic coupling between the spin of the excited electron and the spin of the molecular orbit, referred to as spin-orbit coupling. A static magnetic field has no effect on the electronic spin, but the heavy atom can move and induces a changing magnetic field that leads to a spin flip of the electron. [16]

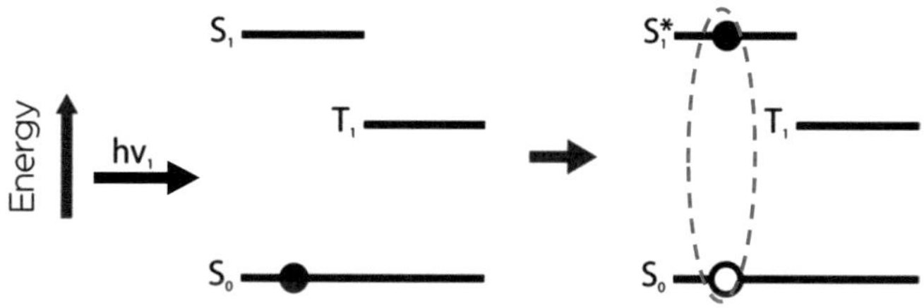

Figure 1.4: From left to right: A photon with energy $h\nu_1$ excites the electron (solid circle) from the ground state S_0 into the higher energy state S_1. After absorption, a hole state (circle) is left in the S_0 state and creates with the excited electron an exciton (red dotted ellipse). An intermediate triplet level T_1 is indicated for completeness. This figure is adapted with permission from Wollgarten's master thesis. [17]

After the creation of an exciton and a possible ISC, the exciton can decay from an energetically higher singlet S_1 or triplet state T_1 to the ground state S_0. This process can be accompanied by the emission of a photon; however, the energy can also dissipate into mechanical degrees of freedom of the molecule, like vibrations or rotations. The radiative decay of a singlet exciton is called fluorescence, whereas the radiative transition from a triplet state to the ground state is called phosphorescence. As phosphorescence requires an additional spin flip, its rate is typically much slower than the rate of the fluorescent decay. In the case of fluorescence, the singlet lifetime can range from picoseconds to nanoseconds. In contrast, the typical triplet lifetimes range from microseconds to seconds.

The formalization of an exciton's time dynamics can be done by rate equations. A simple yet instructive example demonstrates the transition of a singlet exciton to the ground state S_0 or the transfer via ISC to the triplet state T_1:

$$S_1 \begin{matrix} \to S_0 \\ \to T_1 \end{matrix}. \tag{1.1}$$

This transition of the excited state S_1 into either the ground state S_0 or the triplet state T_1 can be deformalized by a differential equation that describes the rate of change in concentration of the excited singlet population S_1 (assuming there is no nonradiative decay from S_1 to S_0):

$$\frac{\partial S_1}{\partial t} = -k_{rad} S_1 - k_{\mathrm{ISC}} S_1. \tag{1.2}$$

Here, k_{rad} is the radiative decay rate that is the inverse of the radiative lifetime, and k_{ISC} is the ISC rate. For the ground state S_0 and the triplet state T_1 the following rate equations apply:

$$\frac{\partial S_0}{\partial t} = k_{\mathrm{rad}} S, \tag{1.3}$$

$$\frac{\partial T_1}{\partial t} = k_{\mathrm{ISC}} S. \tag{1.4}$$

We can calculate the fraction of excited states per decay channel by dividing the relevant rate by the sum of all rates to obtain the radiative yield:

$$\phi_{\mathrm{rad}} = \frac{k_{\mathrm{rad}}}{k_{\mathrm{rad}} + k_{\mathrm{ISC}} + k_{\mathrm{nonrad}}}. \tag{1.5}$$

Here, the non-radiative rate was included for completeness. Besides monomolecular decay processes, excitons can collide and annihilate. Exciton-exciton annihilation can either be promoted by cross-relaxation, meaning that the electron of one molecule relaxes to the ground state of the other molecule. Another possibility is the excitation of one electron to a higher state by the simultaneous relaxation of the second electron into the ground state. The annihilation of excitons involves the collision of two excited states and therefore depends on the square of the excited state concentrations:

$$\frac{\partial S}{\partial t} = -\gamma S^2. \tag{1.6}$$

Here, γS^2 is the collision frequency and is a measure for the rate of collisions in a given volume in a given time.

1.2.1 Exciton transport

Excitons can be mobile, that is they are not confined to the molecule they have been created on but instead they can move through the material. This exciton or energy transfer from one molecule to another is described by two mechanisms. The first involves the exchange of a virtual photon and is due to a Coulomb coupling between the dipole moments of the molecules. This Coulomb exchange mechanism is described 1948 by Förster and involved the experimental measurement of the energy transfer from fluorescin molecules to chlorophyll-a molecules.[18] It is commonly referred to as Förster resonance energy transfer (FRET). Using Fermi's golden rule, the transfer rate for a Coulomb interaction between to identical molecules is: [19]

$$k_{\text{FRET}} \propto \frac{\kappa^2 \mu^4}{R^6} J, \tag{1.7}$$

where κ is the orientation factor between the molecular dipoles, μ is the dipole transition moment that is proportional to the radiative quantum efficiency of the molecule, R the distance between the molecules, and J the overlap of absorption and emission spectra of the corresponding molecules. A typical distance for FRET based energy transfer is the Förster radius. The Förster radius is the distance between the donor and acceptor molecule at which the transfer efficiency based on FRET is 50%. For organic dyes the Förster radius can range from 2 – 10 nm. [20]

The second transfer mechanism involves the tunneling of the electron from one molecule to another due to the overlap of the molecular wavefunctions. This transfer mechanism is a result of the simultaneous tunneling process of the excited electron hole pair between two molecules and is described by

Dexter in 1953. [21] It is commonly referred to as Dexter energy transfer (DET). The transfer rate can again be derived from Fermi's golden rule: [19]

$$k_{\mathrm{DET}} \propto \exp\left(-\frac{\Delta}{R}\right), \tag{1.8}$$

here Δ is the spatial extent of the wavefunction overlap between the involved two molecules. The power law makes the FRET a long-range interaction, but DET a short-range interaction process as it depends exponentially on the distance. A typical interaction radius of DET is ~ 1 nm, meaning that a triplet can only transfer from one molecule to another if they collide or approach each other very closely. Moreover, in solid systems the DET relies heavily on phonon interaction described by the Marcus theory as the molecules cannot move freely. [22] Besides the different distance dependence, the fact that a dipole-dipole interaction involves the exchange of a virtual photon, FRET allows only singlet exciton transfer but not triplet exciton transfer. This is because a FRET of a triplet state would mean a relaxation to the singlet ground state plus the excitation of a new triplet state, both would violate the transition rules for the absorption of a photon. DET is possible for both spin states. Interesting consequents of DET is that organic materials with efficient DET are also suitable for electric conductivity and vice versa. [23] As excitons have a net-charge of zero, they are unaffected by external electic fields. Their movement is instead governed by random hopping from molecule to molecule and can be described as a diffusion process. With this it is possible to define a diffusion length L_D that describes the average distance an exciton can move until it decays.

In general, the diffusion length L_D, the diffusion constant D and the lifetime τ have the following relationship:

$$L_D = \sqrt{nD\tau}, \tag{1.9}$$

where n is 2, 4 or 6 in 1, 2 or 3 dimensions. The diffusion constant D depends on the system. In the case of molecules that are in solution, the diffusion constant of excitons corresponds to the diffusion constant of the specific molecule in a solution. Therefore, the exciton transfer is diffusion limited. In contrast to this, the diffusion of excitons in a solid-state system in which the organic molecules are static is limited to the parameters described in equations (1.7) and (1.8). That is, the orientation, the distance and the dipole moment.

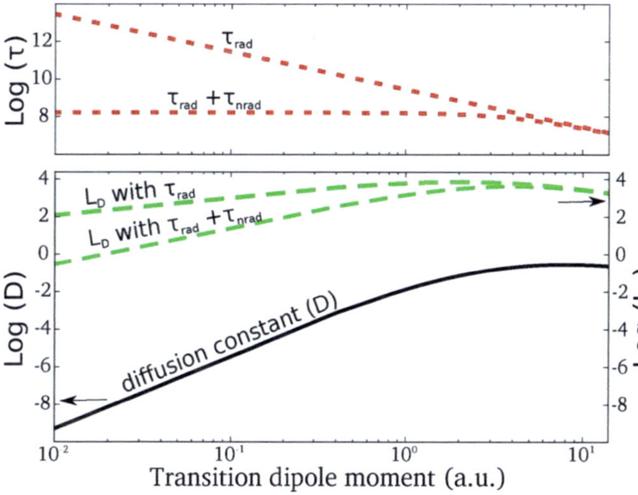

Figure 1.5: The dependence of the lifetime τ, diffusion constant D and the diffusion length L_D is shown as a function of the transition dipole moment μ for excitons moving by Förster resonance energy transfer. Copied with permission from Yost *et al.* [19]

The analysis of the diffusion length of excitons is important for the design of any device based on a specific material. In the case of a bulk heterojunction based organic solar cells, an exciton that is created by the absorption of sun light must travel to the interface between an electron donor and an electron acceptor material. At this interface the exciton dissociates, and free charges are created, in this case, the diffusion length limits the spatial extension of the electron donor and acceptor material. There are several techniques to estimate a diffusion length for excitons. By varying the spatial distribution of quenching molecules the diffusion of excitons length can be estimated. [24] Another method is to introduce a quenching layer at different positions of an active layer within an organic light emitting diode or an organic solar cell. [25] More intrinsic estimations of a diffusion length, that means without the change of the given material system is by performing intensity dependent measurements. The underlying mechanism used is that moving excitons can annihilate. From the fitting of a model to the intensity dependent time dynamics a value for the collision frequency γ can be found. The collision frequency is proportional to the diffusion constant and together with the lifetime of the excitons a diffusion length can be calculated.

Yost et al. estimated an upper limit for the diffusion of singlet excitons. The FRET rate is high if the dipole transition moment μ is high. The dipole transition moment. However, is directly proportional to the decay rate of the singlet exciton. That means that a large μ corresponds to a short lifetime and this shortens the diffusion length. [19] Figure 1.5 shows the lifetime τ, the diffusion constant D and the diffusion length L_D as a function of the dipole transition moment. From this a maximum diffusion length of 200 nm could be estimated for singlet excitons. For DET that is responsible for the motion of triplet excitons such a theoretical limit does not exist.

1.3 Energy transfer in metal-organic frameworks

Metal-organic frameworks (MOFs) are a class of crystalline organic/inorganic synthetic materials in which polytopic carboxylic organic linkers are coordinated by metal centers. The immense range of linkers and metal centers makes this a highly versatile material class. Figure 1.6 shows a small variety of different MOFs with the corresponding metal centers (M) and organic linkers (L). By the variation of the linker molecule many of the MOF's properties can be tuned, such as structure, porosity, stiffness, and chemical interaction with guest molecules.

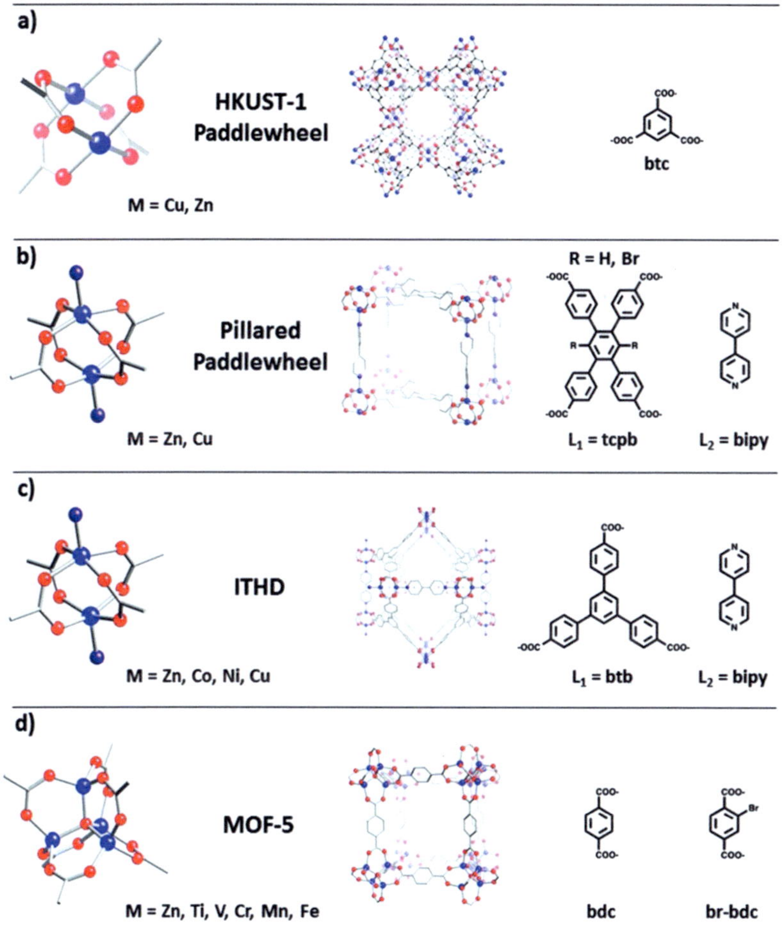

Figure 1.6: Shown are four different metal-organic framework morphologies. The first column shows the configuration of the carboxylic acid groups around the metal centers. The second column shows a corresponding unit cell and the third column possible organic linkers. Copied with permission from Deria *et al.* [32]

This flexibility has driven research targeting applications including gas storage,[26] membrane technologies, [27] and battery materials. [28] Furthermore, a breakthrough that involves the loading of a MOF with appropriate guest molecules enabled a six order of magnitude increase in the electrical conductivity, [29] and such host-guest systems have promising potential for thermoelectrics. [30] Even solar cells were demonstrated, however, with a very low efficiency. [31] The crystalline nature of MOFs makes them interesting model systems for energy transfer processes. The randomness in orientation and distance between chromophores that is found in most other organic semiconductor thin films can be avoided (e.g. in conjugated polymers and thin films of small molecules deposited from solution or by thermal evaporation). Pioneering work on the energy transfer in MOFs has been done in the group of Prof. Wenbin Lin. The first work in this research area has been performed by Kent et al. [33] For the estimation of a diffusion length a mixed MOF made from Ru and Os complexes is synthesized. The energy transfer from Ru to Os is monitored for different ratios between both constituents. From the photoluminescence quenching efficiency of Ru as a function of the Os concentration a diffusion length between 4 nm and 7.5 nm could be estimated.

Following this work, DET is identified as the main energy transfer mechanism in this system as only triplet excitons are involved. [33] Exceptionally long FRET-based diffusion lengths could be measured in Zn(II) porphyrin based MOFs (see Figure 1.7). An average diffusion length of 52 nm is observed. [34, 35] This is further confirmed in a truxene based MOF that is loaded by a quenching molecule (coumarin 343). From the variation of the concentration diffusion lengths up to 42 nm could be measured.[36, 37] The observed diffusion lengths in MOFs are large compared to conventional organic semiconductors that have average diffusion distances of ~10 nm. The main reason for the efficient energy transfer is that the constituents are at close distance and have a fixed orientation. As discussed, both parameters are essential for the transport of excitons and can be tuned by using MOFs.

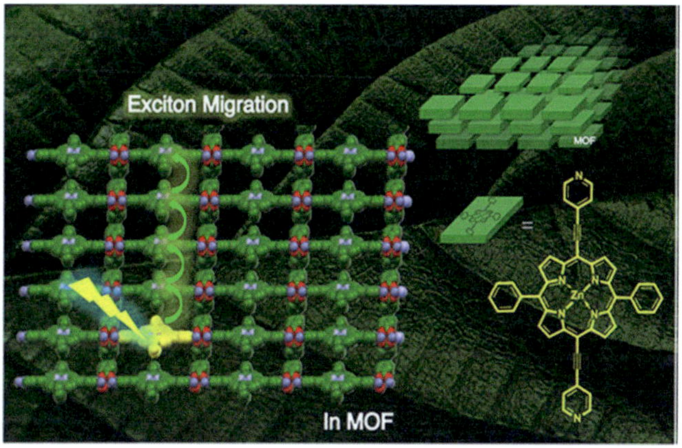

Figure 1.7: The creation of an exciton by illumination (yellow lightning) and the subsequent exciton transport (green arrows) is illustrated a MOF made from Zn-porphyrin linkers. The structure of the linker is shown on the right. Copied with permission from Son *et al.* [38]

1.4 Surface-anchored metal-organic frameworks

The logical implication of good transport in MOFs is to use them TTA-UC. However, MOFs are only available as powders after the synthesis. The deposition of a powder on top of a substrate could theoretically lead to a thin film, however, the orientation of the crystallites would be random. Yet, surface-anchored metal-organic frameworks (SURMOFS) are MOFs that can be deposited as thin films with a preferred orientation.

Surface-anchored metal-organic frameworks (SURMOFs) are a special case of metal-organic frameworks that manifest in thin-films rather than powders. Different then MOFs, SURMOFs are produced by simple spray deposition or spin coating deposition technique. The process for spray coating deposition is sometimes referred to as liquid phase epitaxy (LPE) is illustrated in Figure 1.8. [39] The author would like to state that the term "epitaxy" is Copied with permission from the recent literature and serves solely for a proper reference.

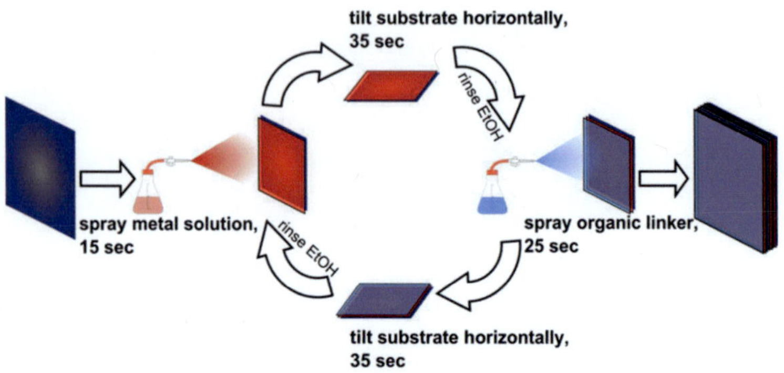

Figure 1.8: The deposition of a surface-anchored metal-organic framework (SURMOF) is illustrated for a flat substrate. EtOH stands for ethanol, that is used as the solvent for both, the organic linker and the metal acetate.

The hydrophilic surface of a substrate is covered by the metal center solution. After washing by ethanol, the linker solution is deposited. The residuals are washed away again, and the metal solution is sprayed on. This process is repeated until the desired thickness is reached. The same repeating approach can be performed by the deposition in a spin coater. [40]

Stavila et al analyzed the kinetics of the growth of the SURMOF HKUST-1 by using a quartz crystal microbalance. [41] HKUST-1 is a prominent example of a SURMOF or MOF in general. The structure is illustrated in Figure 1.6 a). Every step of one cycle could be monitored and it is verified that indeed a layer-by-layer growth is possible. Figure 1.9 shows the increase of mass after each step. The mass increase during the deposition step of the metal center (copper acetate) is smaller than during the deposition of the organic linker (trisomic acid). This is mainly explained by the fact that to each copper acetate molecule three trisomic acid molecules are attached. The growth process is illustrated in Figure 1.10.

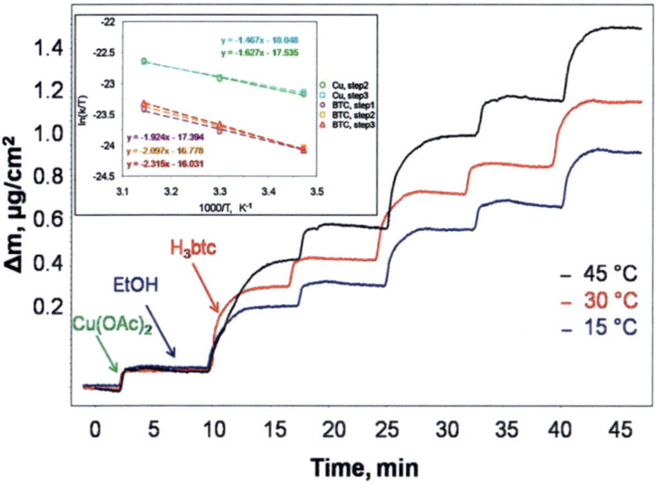

Figure 1.9: Shown is the mass increase after every step of the layer-by-layer deposition for different temperatures. Copied with permission from Stavila *et al.* [41]

However, this idealized monomolecular single layer growth is not the full truth. As is demonstrated by detailed AFM measurements Summerfield et al demonstrated that after each cycle 5-10 layers can be deposited. [42] That is mainly due to non-reacted residuals that stay on the surface even after the rinsing step. Another very interesting characterization by Zhuang et al has shown that the orientation of ditopic carboxylic linker based SURMOFs depends on the temperature.

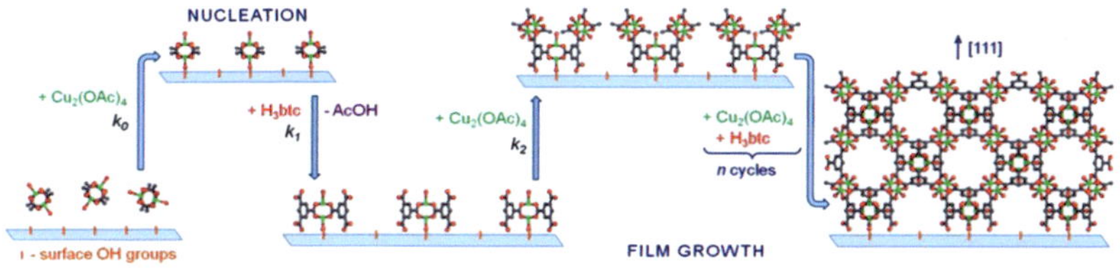

Figure 1.10: From right to left: The OH group terminated surface is covered by the copper acetate molecules (nucleation). Four trisomic acid molecules attach to each cupper acetate. Copied with permission from Stavila *et al.* [41]

At room temperature the orientation of the SURMOF is [100], however, at higher temperature (>60° C) the orientation of the SURMOF is a mixture between [100] and [001]. The case for a SURMOF based on 4,4'-bipyridine and copper acetate is illustrated in Figure 1.11.

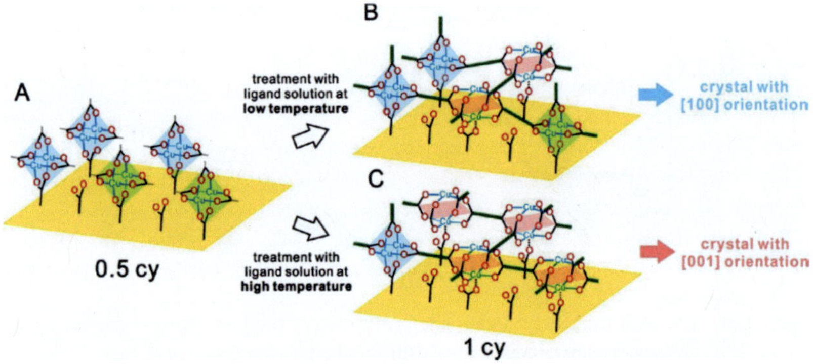

Figure 1.11: A) Shown is the nucleation of copper acetate on the carboxylic group terminated surface. Corresponding to a half cycle. B) In the case of low temperatures during the synthesis shows that the growth orientation is mainly [100]. C) In the case of high temperatures the main orientation is [001]. Copied with permission from Zhuang *et al.* [43]

Beside HKUST-1 new types of SURMOFs are developed in which ditopic carboxylates are used. The use of ditopic carboxylate linkers leads to the formation of SURMOF-2. The name is derived from MOF-2, that is the simplest framework. MOF-2 is based on the paddle-wheel unit formation by attaching 4 carboxylate groups to the metal centers. A selection of SURMOF-2 structures is shown in Figure 1.12. The characteristic of these SURMOFs is that into [001] and [001] direction the distance between the metal centers depends on the length of the used linker, in [010] however the distance is always around ~0.6 nm. [39] The layer-by-layer growth allowed the growth of SURMOF-2 with different unit cell sizes on top of each other. [44]

The first approach towards an optoelectronic device is performed by Liu *et al.* A dye-sensitized solar cell is built which had an efficiency of 0.2%. The linker that is used is a derivative of Pd-Porphyrin and free-base porphyrin. [45] In the mentioned publication the hypothesis is that the photovoltaic action is due to the dissociation of triplet states into free charges making SURMOF-2 unique organic semiconductors.

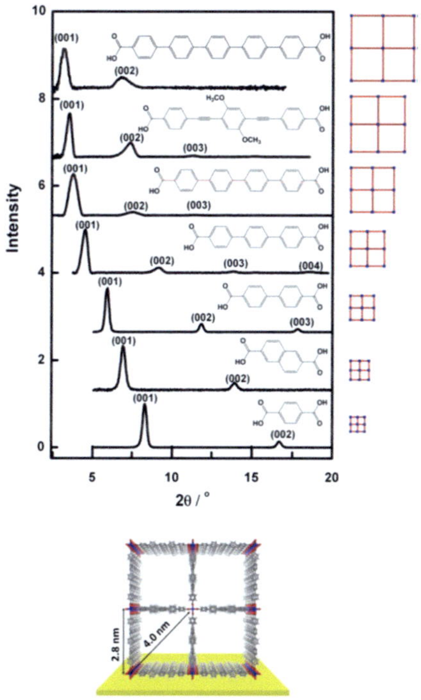

Figure 1.12: Top: Shown is the out-of-plane X-ray diffractogram and the molecular structure of seven different ditopic carboxylate linkers. Bottom: An illustration of four unit cells is shown for a particular linker. Copied with permission from Liu *et al.* [39]

1.5 Organic photon upconversion

As previously mentioned photon upconversion (UC) is a process where the absorption of two low-energy photons results in the emission of one photon with higher energy. Organic photon upconversion involves two kinds of organic molecules. The first molecule is the absorber and is referred to as sensitizer. The second kind of molecule is the emitter. The absorption of the photons by a sensitizer molecule leads to the creation of singlet excitons followed by fast ISC of the singlets to the triplet state. These triplets can transfer onto a second molecule (emitter) where they collide and perform pairwise triplet-triplet annihilation. This process leads to one of the triplets returning to the ground state while the other one is promoted to an excited singlet state and can decay radiatively. The first experimental observation of TTA-UC was made by Parker et al. as an experimental tool to prove that excited molecules undergoing ISC form triplet states that can transfer their energy onto molecules with no ISC. The measured delayed fluorescence was the first demonstration for TTA-UC. [46]

1.5.1 Applications of TTA-UC

A vulnerability of TTA-UC is that the sensitizer is quenched by oxygen before it undergoes ISC and transfers the triplet excited state onto the emitter. Borisov *et al* used the vulnerability in their favor and build an oxygen sensor by using PdTBP as the sensitizer and a perylene derivative as the emitter. The sensor material is illuminated by a light emitting diode (LED) and by adding oxygen the UC emission is quenched. [47]

Another area of application are OLEDs. As charge carriers are injected the bounding with holes can result in one emissive singlet state and three dark triplet states. This leads to a total loss of 75%. However, by using emitter molecules that perform efficient TTA this can be partially overcome. Kondakov *et al*l demonstrated that the efficiency of the red electroluminescence could only be explained by the annihilation of the produced dark triplet states. [48]

The direct utilization of the energy conversion process is demonstrated for the activation of a ruthenium predrug complex by using PdTBP and perylene. Askes *et al* could demonstrate that by shining red light onto the system the predrug could be activated by the blue light and used for photodynamic therapy. [49] Another biological application is demonstrated by Wohnhaas *et al* by using TTA-UC nano capsules for tracing the uptake of cells. [50]

Water splitting that is based on the activation of titanium dioxide by UV light is demonstrated for lower energy light excitation by using TTA-UC. Monguzzi *et al* demonstrated an increase efficiency of water splitting by using PtOEP and DPA and illuminating with 532nm. [51] Further, photo degradation of ethyl violet on a titanium dioxide surface is demonstrated to work under visible light by using the aforementioned TTA-UC system.[52]

TTA-UC has also been investigated for light absorbing materials, as they are used for photovoltaics or photocatalysis. The main idea is to attach an upconverting system behind the absorbing material of the solar cell and reflecting the upconverted light back to active the response of the solar cell for light with energy lower than the bandgap. Yet, this approach suffers from possible optical losses. Lissau *et al* demonstrated that the direct energy transfer to the active material is possible and optical losses can be avoided. DPA is attached to the surface of zirconium oxide micro particles and immersed into a solution with PtOEP. The strong quenching of the upconversion signal is the indication for the transfer of the upconverted energy to the microparticles. [53-55] Hill *et al* optimized this approach by attaching PtOEP molecules to the DPA molecules. This has shown the same quenching of the upconversion emission as in the aforementioned system. [56] The next step is performed by Hanson *et al* by fabricating dye-sensitized solar cells. DPA is coordinated to the titanium dioxide surface and then PdOEP and PtOEP are subsequently attached to the DPA layer. [57]

The suitability of a TTA-UC system for applications depends on three characteristics. The first is the wavelength shift that depends solely on the used molecules. The second characteristic is the UCQY that states the ratio of upconverted photons and absorbed photons. The third characteristic is the upconversion threshold that states under which excitation power the TTA-UC works efficiently. In the next section the former two characteristic quantities, that is referred to as figures of merit of TTA-UC is discussed for different systems and how they are optimized.

1.5.2 Figures of merit in TTA-UC

TTA-UC emission shows a transition from a quadratic to a linear dependence on the excitation power density. The TTA-UC quantum yield shows therefore a transition from linear to a constant value (as is shown in Figure 1.12). This behaviour allows the definition of two figures of merit that allow the performance to be compared of TTA-UC in different molecular solutions and in different molecular arrangements. The first one is the maximum internal UCQY, which is the multiplication of the efficiencies of each step from absorption to the emission of an upconverted photon:

$$\phi_{UC} = 1/2\, f \phi_{ISC} \phi_{ET} \phi_{TTA} \phi_{PLQY}. \tag{1.10}$$

where ϕ_{ISC} is the efficiency for ISC, ϕ_{ET} is the energy transfer efficiency from sensitizer to emitter, ϕ_{TTA} is the TTA efficiency and ϕ_{PLQY} is the photoluminescence quantum yield (PLQY) of the emitter. The factor of ½ describes the fact that two absorbed photons are required to generate one upconverted photon, and f represents the spin-statistical factor describing how many TTA events result in a singlet state. The TTA efficiency is intensity dependent as it requires the collision of two excited states. The maximum upconversion quantum efficiency is therefore reached in the regime of high triplet densities when ϕ_{TTA} asymptotically approaches unity. The intensity dependence of the quantum efficiency and the first figure of merit, the maximum upconversion efficiency UC-$PLQY_{max}$ are shown in Figure 1.13A.

The second important figure of merit for photon UC systems is the power density of excitation light necessary for the bimolecular TTA-UC to become as efficient as the monomolecular triplet decay. This figure of merit has been named the *UC threshold.* This value can be estimated by considering the dependence of UC intensity on the excitation power density.

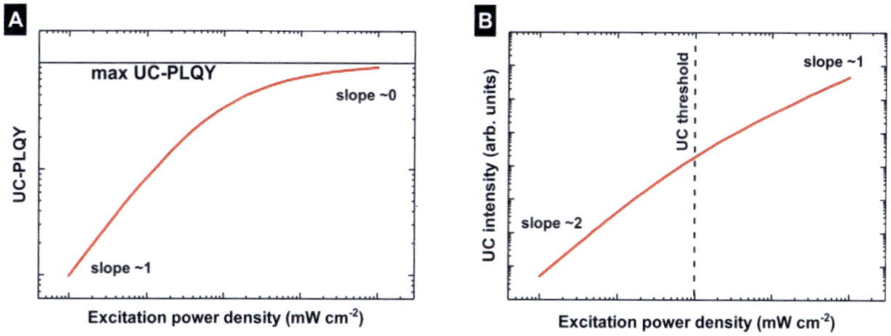

Figure 1.13: A) UC-PLQY as a function of the excitation power density. The horizontal line marks the saturation value for UC-PLQY. B) UC intensity as a function of the excitation power density. The dashed line marks the upconversion threshold, which can be estimated from the asymptotes at very low and very high excitation power density.

In Figure 1.13B, two distinct regimes can be observed: For low power densities the slope of the UC intensity curve exhibits a quadratic behavior, for high power densities the slope is linear, which means that the triplet density is sufficiently high that the bimolecular TTA rate is much faster than the monomolecular loss and any additionally absorbed pair of photons leads to the emission of an upconverted photon. The UC threshold can be defined as the transition point between the two regimes, the point at which the monomolecular decay rate and the rate of TTA become equal (this is analyzed more details in the upcoming results chapter). This is usually done by fitting a linear and quadratic function to the high- and low-power regions respectively and determining the intersection of the two lines as is demonstrated in Figure 1.13B.

1.5.3 State-of-the-art TTA-UC systems

Equation (1.10) describes the upconversion quantum efficiency and suggests that the optimization of each factor results in high upconversion efficiency.

The first limiting quantity is the spin statistical factor f. There are nine possible results of a TTA event: one leads to the generation of a singlet complex, three to the generation of a triplet complex and five to the generation of a quintet complex. [58] Assuming that the probability for each state is the same, the probability for the creation of a singlet state is f=1/9. Assuming that the other efficiencies are close to unity gives an upper limit for the TTA-UC quantum efficiency of 5.5%. The spin statistic factor f is analyzed for various emitter molecules. A system utilizing rubrene as the emitter and Pd(II) tetrakisquinoxalinoporphyrin showed a maximum TTA-UC efficiency of 8%, which is higher than the value predicted by spin statistics. This observation shows that the energy level of the quintet and/or triplet states lies higher than the singlet state of rubrene and cannot be accessed after TTA. [58] A later analysis of Pd(II) octaethylporphyrin (PdOEP) with anthracene showed that the quintet channel is totally quenched and demonstrated 20% of TTA efficiency. [59] Perylene is found to be an ideal emitter material for TTA-UC, showing an external UCQY approaching the ideal value of 50%. This indicates that the spin the statistic factor f can be close to unity if the emitter is chosen so that the triplet and quintet products of exciton-exciton annihilation are energetically inaccessible.[13]

The second limiting process is ISC. Islangulov *et al.* demonstrated TTA-UC performing at visible wavelengths by using the ruthenium complex as a sensitizer molecule together with diphenylanthracene (DPA) and anthracene as the emitter molecule. The motivation behind using Ru(dmb)(3) is that ruthenium exhibits efficient ISC as the heavy atom introduces strong spin-orbit coupling. [60] This is later surpassed by porphyrin molecules with host transition-metal atoms like palladium or platinum. [61-63] At the current stage of research porphyrin molecules are leading the benchmark for sensitizer molecules in TTA-UC as there ISC efficiency has been observed to be close to unity. The third limitation arises from the transfer of the sensitizer triplet to the emitter molecule. Throughout the literature, energy transfer efficiencies close to unity are achieved for porphyrin sensitizer molecules at sufficiently low sensitizer/emitter concentration ratios. Finally, the last efficiency in Equation 1.9 is the TTA efficiency. At sufficiently high excitation power densities ϕ_{TTA} reaches unity. For a systematic comparison between different TTA-UC systems Monguzzi *et al.* introduced the upconversion excitation threshold: [64]

$$I_{TH} = \frac{1}{\alpha}\frac{1}{\phi_{ET}}\frac{{k_E^D}^2}{\gamma}. \tag{1.11}$$

Here, I_{TH} has the units of photon flux, α is the fraction of photons absorbed divided by the beam length, k_E^D is the emitter triplet decay rate, and γ is the collision frequency. The collision frequency γ is proportional to the diffusion constant of the molecules in a given environment. Hence, ϕ_{TTA} is a diffusion-limited process and it is assumed that the increase of the emitter concentration will lead to more frequent collisions. [65] Yet, an increase of the emitter concentrations to reduce I_{TH} is shown to lead to the quenching of ϕ_{PLQY} by the formation of non-radiative excimer states in a system made from [Ru(dmb)3] and 9,10-dimethylanthracene.[66] However, by using ionic liquids the concentration of perylene as an emitter could be increased without showing aggregation losses. [67-70] As a significant example of an ionic liquid system Pt(II) octaethylporphyrin (PtOEP) together with DPA shows a low threshold of $I_{TH} = 3\,mW/cm^2$ and a quantum yield of $\phi_{UC} = 5.6\%$. [71] For the implementation into a device the external quantum yield plays a significant role. The external quantum yield is the internal quantum yield ϕ_{UC} multiplied by the fraction of photons absorbed n:

$$\phi_{UC}^{ext} = n \cdot \phi_{UC}^{int}. \tag{1.12}$$

Concentration-dependent measurements have been performed by using Pd-octaethylporphyrin (PdOEP) and DPA, showing that an external UCQY of 1% is possible. The experimental data shows that the best performance is achieved with a sufficiently low sensitizer concentration of 1%. [61] Further enhancement of absorptivity and energy transfer is achieved by using multiple emitter or sensitizer molecules. An interesting system with two emitter molecules is demonstrated by Turshatov *et al*. The external UCQY is found to be 11% for the mixed system. In comparison, the external UCQY for a single emitter system is around 6%. The explanation for this enhancement is that the broader distribution of triplet levels allows better energy transfer efficiencies. [72] By using six different porphyrin derivates as sensitizer, Monguzzi *et al* could cover a big range of the solar spectrum and demonstrating an external UCQY up to 10%. [73, 74]

1.5.4 State-of-the-art TTA-UC for clustered emitter molecule systems

The TTA-UC systems presented in the previous section are either based on solutions or on polymer matrix. The random distribution of emitter molecules and sensitizer molecules in both systems is making an efficient energy transport difficult. The energy transport however, is significant for a low threshold and a high quantum yield and so for the application of TTA-UC. A new route for TTA-UC systems that paves the way for applications are systems that cluster emitter molecules so that these form agglomerates of molecules. The resulting close distance between the emitter molecules results in good energy transport.

Systems made from covalently bound emitter molecules are considered already very early. However, only recently efficient systems are developed. The first demonstration of TTA-UC within a covalently bound emitter system is demonstrated by loading a polyfluorene matrix with PtOEP. In this system, the polyfluorene matrix itself acts as the emitter. Despite the very low efficiency and the high excitation intensities needed to detect UC, this is the first demonstration of an emitter system in which the emitter molecules are close together allowing efficient energy transfer. [75] This work is later followed up by Baluschev *et al.* using different kinds of coordinated emitter systems doped with PtOEP.[76-78] However, the used emitters had high triplet levels that consequently lowered ϕ_{ET} and with this the UC-PLQY significantly. Another very similar approach is performed by Jankus *et al.* TTA-UC is demonstrated in a poly(para-phenylene vinylene) super yellow copolymer which is doped with Pd(II)-Tetrabenzoporphyrin (PdTBP). Although the copolymer has lower lying triplet levels the system is still inefficient. The main loss in this system comes from strong aggregation formation of the sensitizer PdTBP.[79] Another concept is to use PdOEP within a DPA based copolymer. Due to aggregation effects the UC signal is small but the threshold is comparably low $I_{TH} = 32\, mW/cm^2$. [80]

Another approach is to make systems in which the emitter molecules are manipulated in such a way, that they interact strongly with each other. This strong interaction leads to a self-assembly of the emitter molecules. The first successful system of this kind is demonstrated by Ogawa et al. using a derivative of DPA. The molecular structure allowed it to self-assemble while the distance between emitter molecules is kept sufficiently high and so both the threshold and the internal UCQY could be enhanced. The arrangement of DPA in close proximity lead to a threshold of $I_{TH} = 8.9mW/cm^2$. In addition, the oxygen protective characteristics of the emitter self-assembled molecular system resulted in a high internal UCQY of 15%. [81] The diffusion constant for the emitter triplets is estimated to be $D = 1.4 \cdot 10^{-5}\, cm^2/s$. That is close to diffusion constants found in anthracene crystals $D = 1.2 \cdot 10^{-5}\, cm^2\, s^{-1}$. [82] Furthermore, a gel made from a derivative of DPA has shown a remarkably small threshold of only $I_{TH} = 1.5\, mW\, cm^{-2}$. [83]

1.5.5 TTA-UC in ordered molecular systems

As previously mentioned in TTA-UC there are two figures of merit that determine the performance of the system. For applications UC-PLQY$_{max}$ has to be maximized and the threshold I_{TH} minimized. The upconversion threshold depends on the mobility of the excitons, which means that if it the mobility can be increased the threshold should get lower as well. This is shown for certain TTA-UC in which the emitter molecules are coordinated and doped with sensitizer molecules. The static arrangement between the emitter molecules allowed excited states to move fast from molecule to molecule. However, in these systems the emitter molecules are still randomly arranged leading to a non-optimized energy transfer that also depends on the orientation of the dyes. In contrast, MOFs have been shown to exhibit high diffusion lengths for excited states. That leads to the logical conclusion, that TTA-UC within MOFs should show low thresholds. Further, the threshold depends also on the amount of triplet excitons that are transferred from the sensitizer to the emitter. This has not been demonstrated, yet. Zimmermann *et al.* presented in a theoretical framework that the clustering of sensitizer and emitter molecules is leading to higher efficiencies and annihilation rates compared to a system in which sensitizer and emitter molecules are randomly distributed. [84] The main conclusion is that due to the clustering of sensitizer molecules the local absorption is increased and so the local triplet density. The high local triplet density leads then to a higher collision rate compared to a system in which the absorption is uniformly distributed.

SURMOFs offer a platform to demonstrate this geometric dependency by building layered structures. The main approach is to grow a thick sensitizer SURMOF that absorbs a big fraction of incident light. The created triplet excitons transfer onto a second SURMOF layer made from emitter molecules. To increase the probability of collision in the emitter SURMOF, the thickness of the emitter layer should be kept smaller than the diffusion length of the emitter triplets.

1.6 Outline

The aim of research groups working on TTA-UC is to develop a system that has high quantum yields, low thresholds and can be implemented into a device structure. SURMOFs offer the possibility to fulfill these requirements.

Following this introduction, **Chapter 2** introduces the used experimental and theoretical methods to obtain the main results of this work.

Chapter 3 presents the standard equation system for TTA-UC. Under certain approximations the steady-state and time dependent solutions for the excited state concentrations are derived. At the end, the hypothesis for a bilayer TTA-UC system is derived, which paves the way for the next chapters. The results of this chapter will be published:

M. Oldenburg, R. E. Joseph, N. Kiseleva, D. Kim, D. Busko, B. S. Richards, I. A. Howard, *Triplet-Triplet Annihilation Upconversion Threshold Determination: A Comparison of Steady-State and Time-Resolved Methods,* in preparation.

The derived master curve is published and used for characterization in:

N. Kiseleva, M. A. Filatov, **M. Oldenburg**, D. Busko, M. Jakoby, I. A. Howard, B. S. Richards, M. Senge, S. M. Borisov, A. Turshatov, The Janus-Faced Chromophore: A Donor-Acceptor Dyad with Dual Performance in Photon Up-conversion. Chem. Commun., 2018, Accepted Manuscript.

Chapter 4 shows the characteristic of the emitter SURMOF layer made from a dicarboxylic derivative of DPA. Strategies for the enhancement of the PLQY are presented and used to understand the exciton dynamics in this system. This chapter builds on the following publication:

M. Oldenburg, A. Turshatov, D. Busko, M. Jakoby, R. Haldar, K. Chen, G. Emandi, M. O. Senge, C. Wöll, J. M. Hodgkiss, B. S. Richards and I. A. Howard, *Enhancing the photoluminescence of surface anchored metal–organic* *frameworks: mixed linkers and efficient acceptors,* Phys. Chem. Chem. Phys., 2018, DOI: 10.1039/c7cp08452h.

Chapter 5 deals with the sensitizer SURMOF layer made from a Pd(II)-porphyrin dicarboxylic linker. The important parameters ϕ_{ISC} and the triplet lifetime is measured. Parts of this chapter will be published in:

M. Adams, N. Baroni, **M. Oldenburg**, F. Kraffert, R. W. MacQueen, D. Busko, A. Turshatov, G. Emandi, M. O. Senge, C. Wöll, K. Lips, B. S. Richards, I. A. Howard, *Charge photogeneration through the triplet state in porphyrin-based surface-anchored metal-organic frameworks,* in preparation.

Chapter 6 introduces the bilayer approach and discusses the limitations. This chapter is based on the following publication:

M. Oldenburg, A. Turshatov, D. Busko, S. Wollgarten, M. Adams, N. Baroni, A. Welle, E. Redel, C. Wöll, B. S. Richards and I.A. Howard, *Photon Upconversion at Crystalline Organic–Organic Heterojunctions.* Adv. Mater., 2016, 28: 8477–8482.

M. Oldenburg, A. Turshatov, D. Busko, M. Jakoby, R. Haldar, K. Chen, G. Emandi, M. O. Senge, C. Wöll, J. M. Hodgkiss, B. S. Richards and I. A. Howard, *Enhancing the photoluminescence of surface anchored metal–organic frameworks: mixed linkers and efficient acceptors,* Phys. Chem. Chem. Phys., 2018, DOI: 10.1039/c7cp08452h.

Chapter 7 serves as a conclusion and shows ways for further developments.

2 Methods

This chapter serves as a descriptive chapter for the used theoretical and experimental methods, that are used to obtain the results of this thesis. This chapter is divided into three parts. The first part describes how to solve special differential equations that are used to describe the time dynamics of excitons. The solutions will help to extract certain parameters. The second part describes the experimental spectroscopic methods including the measurement of the absorption spectrum, the PLQY, the streak camera that was used for the time resolved PL and the high sensitivity setup that is used to measure very weak signals. The third part describes the synthesis of SURMOFs and how they are characterized.

2.1 Theoretical methods

2.1.1 Selection of special differential equations

As described in the introduction chapter, the time dependence of excitons in a given material can be described by rate equations. In the following the rising edge for excitons that have a mono- and a bimolecular decay channel are analyzed under a constant excitation. In the second part, the falling edge of the corresponding system is analyzed. A fitting of the found functions to experimental data can provide important insights into the system.

Time dynamics of the rising edge for excitons with a mono- and a bimolecular decay channel

The rising edge of at a constant excitation can be described by using the following rate equation:

$$\frac{\partial E_T}{\partial t} = G - (k_E^D + \gamma E_T)E_T. \tag{2.1}$$

Here G is the generation rate of the excitons (which is assumed to be a constant), k_E^D is the monomolecular decay rate and γE is the bimolecular decay rate. By multiplying this equation with $\gamma / {k_E^D}^2$ and performing following transformations:

$$\frac{\gamma E_T}{k_E^D} \to \tilde{E}, k_E^D t \to \tau_E, \tag{2.2}$$

the rate equation gets dimensionless.

$$\frac{\partial \tilde{E}}{\partial \tau_E} = \tilde{G} - \left(1 + \tilde{E}\right)\tilde{E}. \tag{2.3}$$

The initial value condition is chosen to be $\tilde{E}(\tau_E = 0) = 0$, which means that there are no excitations in the beginning. This is a special type of a Riccati differential equation. By using the following transformation:

$$\tilde{E}_T(\tau_E) = -\frac{z'(\tau_E)}{z(\tau_E)}, \tag{2.4}$$

the non-linear first-order differential equation (2.3) can be simplified to a linear second-order differential equation:

$$z'' + z' - \tilde{G}z = 0. \tag{2.5}$$

This differential equation can be solved by using the standard approach for linear differential-equations $z = a \cdot \exp(-\lambda \tau_E)$. Leading to a characteristic equation for λ:

$$\lambda^2 - \lambda - \tilde{G} = 0, \tag{2.6}$$

with the solutions:

$$\lambda_{1/2} = \frac{1}{2}\left(1 \pm \sqrt{1 + 4\tilde{G}}\right). \tag{2.7}$$

The found solution can be inserted into equation (2.4):

$$\tilde{E} = \frac{a_1\lambda_1 + a_2\lambda_2 \exp(-(\lambda_2 - \lambda_1)\tau_E)}{a_1 + a_2 \exp(-(\lambda_2 - \lambda_1)\tau_E)}. \tag{2.8}$$

The initial value condition $\tilde{E}(\tau_E = 0) = 0$ leads to the following relation:

$$\frac{a_2}{a_1} = -\frac{\lambda_1}{\lambda_2}. \tag{2.9}$$

And so:

$$\tilde{E} = \frac{2\left(1 - \exp\left(\sqrt{1 + 4\tilde{G}}\tau_E\right)\right)\tilde{G}}{1 - \sqrt{1 + 4\tilde{G}} - \left(1 + \sqrt{1 + 4\tilde{G}}\right)\exp\left(\sqrt{1 + 4x}\tau_E\right)}. \tag{2.10}$$

This can be further rewritten to:

$$\tilde{E} = \frac{2\tilde{G}}{1 + \sqrt{1 + 4\tilde{G}} \coth\left(\sqrt{1 + 4\tilde{G}}\frac{\tau_E}{2}\right)}. \tag{2.11}$$

Time dynamics of the falling edge for excitons with a mono- and a bimolecular decay channel

The rate equation describing the time dynamics of $\tilde{E}$ after excitation is following equation (2.3). However, it was assumed that γ is constant in time for a more general discussion it will be assumed that γ is time dependent. This difference is significant for systems in which excitons are restricted to two or one dimensions. For generality, we assume however a time-dependent collision frequency $\gamma \rightarrow \gamma(t)$. The initial concentration of excitons can be calculated from equation (2.11) and is referred to as E_0.

$$\frac{\partial E}{\partial t} = -k_E^D E - \gamma(t) E^2, E(0) = E_0 \tag{2.12}$$

This special case of the Bernoulli differential equation which can be transformed into a linear differential equation by using:

$$E^{\star} = \frac{1}{E} \tag{2.13}$$

The resulting differential equation for $E^{\star}$ is:

$$\frac{\partial E^{\star}}{\partial t} = k_E^D E^{\star} - \gamma(t). \tag{2.14}$$

The collision frequency is time-independent for the three-dimensional case and the solution of equation (2.14) is:

$$E^{\star} = \frac{\gamma}{k_E^D}(1 - \exp(-k_E^D t)) + E_0^{\star} \exp(k_E^D t) \tag{2.15}$$

The back-transformation $E \rightarrow 1/\tilde{E}$ leads to:

$$E = \frac{1}{E^{\star}} = \frac{k_E^D E_0}{\exp(k_E^D t)\,(k_E^D + \gamma E_0) - \gamma E_0} \tag{2.16}$$

Using $\tilde{E} = \gamma E / k_E^{D^2}$ and $\tau_E = k_E^D t$ as in the previous section leads to the time dynamics for the falling edge of $\tilde{E}$:

$$\tilde{E} = \frac{1}{\frac{1}{\phi_{TTA}} \exp(\tau_E) - 1}, \tag{2.17}$$

with $\phi_{TTA} = \gamma\tilde{E}(\tau_E = 0)/\left(k + \gamma\tilde{E}(\tau_E = 0)\right)$ as the efficiency for the bimolecular decay rate directly before the turn-off of the excitation source.

For certain measurements the integration of equation (2.16) plays an important role and can be performed by the following substitution $x = \exp(-kt)$:

$$\int_0^\infty E(t)\,dt = \frac{k}{\gamma}\ln\left(1 + \frac{\gamma}{k}E_0\right) \tag{2.18}$$

In the case in which excitation are restricted to the movement only in one dimension the collision frequency is time-dependent $\gamma(t) = 4\pi DR_0 \frac{R_0}{\sqrt{2\pi Dt}} = \alpha/\sqrt{t}$. Using the same transformation as before, however, this time with a time-dependent gamma the following equation can be obtained:

$$\frac{\partial E^\star}{\partial t} = kE^\star - \frac{\alpha}{\sqrt{t}}, E^\star(0) = E_0^\star \tag{2.19}$$

For large times this function should have the form $\tilde{E}(t) = \exp(kt)$, therefore we guess the following general solution:

$$E^\star = g(t)\exp(kt). \tag{2.20}$$

Inserting this into the equation (2.19) leads to a differential equation for $g(t)$:

$$g'(t) = -\frac{\alpha}{\sqrt{t}}\exp(kt) \tag{2.21}$$

This can be integrated using the substitution $t' = -u^2$:

$$g(t) = \sqrt{\frac{\pi}{k}}\alpha\,\mathrm{erf}\left(\sqrt{kt}\right) + c \tag{2.22}$$

so that a solution for equation (2.19) is found:

$$E^{\star} = \left(\sqrt{\frac{\pi}{k}}\alpha\,\mathrm{erf}\left(\sqrt{kt}\right) + c\right)\exp(kt), \tag{2.23}$$

c is a constant, which can be found using the initial condition $E^{\star}(0) = E_0^{\star}$.

The backtransformation of $E^{\star}(t)$ to $E(t)$ gives the final solution:

$$E(t) = \frac{1}{E^{\star}} = \frac{E_0\exp(-kt)}{E_0\sqrt{\frac{\pi}{k}}\alpha\,\mathrm{erf}\left(\sqrt{kt}\right) + 1} \tag{2.24}$$

The found solutions for the rising and the falling edge for excitons that have a mono- and a bimolecular decay rate will serve as fitting curves for later measurements.

2.2 Experimental methods – Spectroscopy

The spectroscopic analysis serves as a characterization of excitons. In the following, steady-state spectroscopic methods for measuring the absorption spectrum and the emission spectrum will be shown and explained.

2.2.1 Absorption spectra

Light that falls onto a material is either reflected, transmitted or absorbed. The absorption spectrum tells something about the ground state absorption in a material. For the measurement of the absorption spectrum a spectrophotometer by Perkin Elmer is used (Perkin Elmer 950). The absorption spectrum of SURMOFs that are solid films are measured in the integration sphere in the Perkin-Elmer 950 spectrophotometer. Solutions are measured by placing a cuvette with dissolved dye into the beam path and a cuvette purely with the solvent into the reference beam. In both measurements the absorption spectrum is obtained from the difference in spectrum between the reference and the measurement beam gives the absorption spectrum. The measurement inside the integration sphere reduces parasitic reflections.

2.2.2 Calibration of detectors for PL spectra measurements

The sensitivity of detectors for a given wavelength is governed by the quantum yield at a given wavelength. Further, the measured signal of a detector is in units of an electric current or voltage. Therefore, to calibrate the detector's wavelength sensitivity and for physical signal interpretation a calibration is needed. The calibration lamp that is used for all photoluminescence spectra is the HL-3 plus-CAL-EXT from Ocean Optics which is a tungsten halogen source. By using a tungsten halogen source, the emitted spectrum can be assumed to be a black body spectrum with the corresponding temperature. Figure 2.1 shows the measured spectrum of a calibration lamp by using the Hamamatsu R928P photomultiplier tube (PMT), the correct spectrum (corresponding to a black body curve) and the correction factor. The correction factor results from the division of the black body curve and the measured curve. By performing the calibration any PL signal with arbitrary units as intensity can be transformed into a photon flux, which is important for the measurement of PLQY.

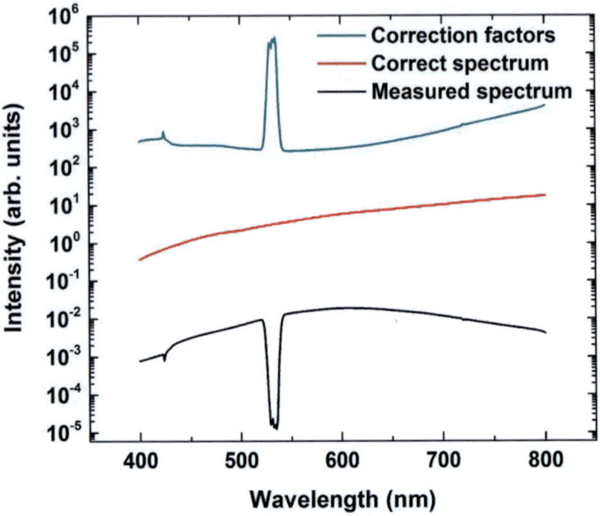

Figure 2.1: Black: The spectrum of the calibration lamp HL-3 plus-CAL-EXT is shown for the wavelength range 400 – 800nm. The dip in the measured curve comes from a 532 nm notch filter that is put into the beam path. Red: The correct black body spectrum of the lamp. Blue: The correction factor.

2.2.3 Photoluminescence quantum yield measurements

The PLQY is the ratio of photons emitted by photons absorbed:

$$\phi_{PLQY} = \frac{n_{\gamma,emitted}}{n_{\gamma,absorbed}}. \tag{2.25}$$

The PLQY gives the efficiency of a material to convert the absorbed photons into photons again. In the following the principle of the measurement is described and after this more detail on the used detectors and excitation sources will be given.

Measurement principle

Figure 2.2 shows the scheme for the measurement of the PLQY introduced by de Mello *et al.* [85] Figure 2.2 A shows the integration sphere with the excitation source and no sample. The integration of the spectrum over the wavelength gives the total photon flux L_A. Figure 2.2B shows the direct excitation of the sample. The spectrum consists of two parts, i) the excitation source intensity L_B that is reduced due to absorption A and ii) the PL spectrum of the sample P_B. Figure 2.2C shows the case of indirect excitation, that is when the sample is not in the beam path of the excitation source.

The spectrum shows again two parts, i) the excitation spectrum L_C reduced by the indirect absorption μ and ii) the PL spectrum that comes from the indirect absorption P_C. Taking indirect excitation into account following three equations can be formulated:

$$L_B = L_A(1-\mu). \tag{2.26}$$

$$L_C = L_A(1-A)(1-\mu) \tag{2.27}$$

$$A = 1 - \frac{L_C}{L_B} \tag{2.28}$$

The overall intensity measured in the direct excitation experiment is:

$$L_C + P_C = (1-A)(L_B + P_B) + \phi_{PLQY} L_A A, \tag{2.29}$$

where the first contribution comes from laser light scattered in the integration sphere and the second contribution from collimated laser light. The quantum yield measured in this experiment is therefore:

$$\phi_{PLQY} = \frac{P_C - (1-A)P_B}{L_A A} \tag{2.30}$$

In the case of weak PL from the sample a long integration time is needed. However, this can lead to an oversaturation of the excitation source signal. The situation is illustrated in Figure 2.3A. The oversaturation leads to an electronic coupling between the pixels of the detector, this influences the spectrum of the sample.

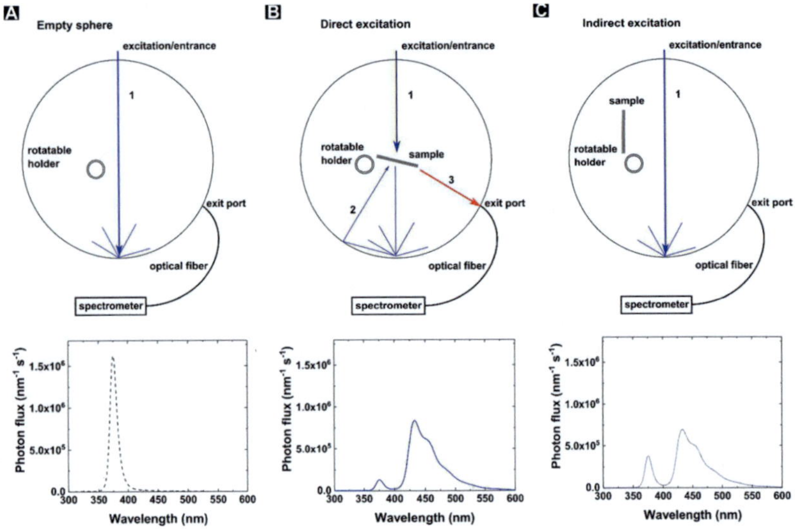

Figure 2.2: A) Showing the integration sphere without sample and the excitation source spectrum. B) Shows the case with direct sample illumination. Indicating the indirect excitation by the scattered laser light 2. C) Shows the situation in which the sample is not in the beam path and is only illuminated by the scattered laser light.

For correction, the P background that is the spectrum of the pure excitation source with the same integration time as used for direct and indirect excitation is measured. The subtraction of the P background leads to a pure PL spectrum shown in Figure 2.3C.

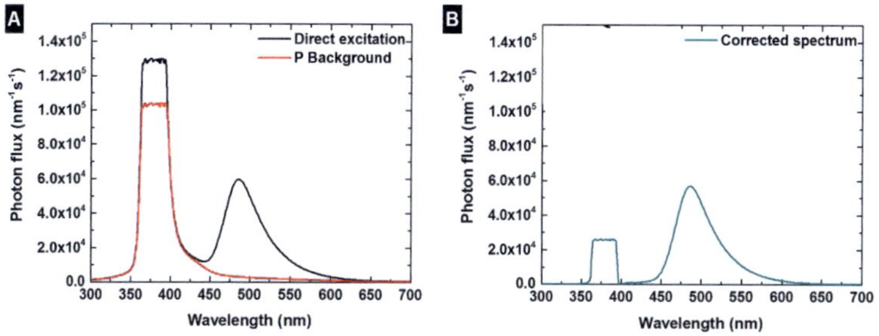

Figure 2.3: A) Spectrum with saturated excitation peak and PL from the sample (black). The spectrum of the excitation source (red) multiplied by an appropriate factor. C) Corrected spectrum after the subtraction of the P background.

In the case of strong absorption and small stokes shift, the sample's PL can be a source of excitation as well. This leads to a red-shift of the PL spectrum if $\phi_{PLQY} < 1$. Figure 2.4 demonstrates this situation. The spectrum of one sample is recorded outside and inside the integration sphere, later red shifted due to reabsorption.

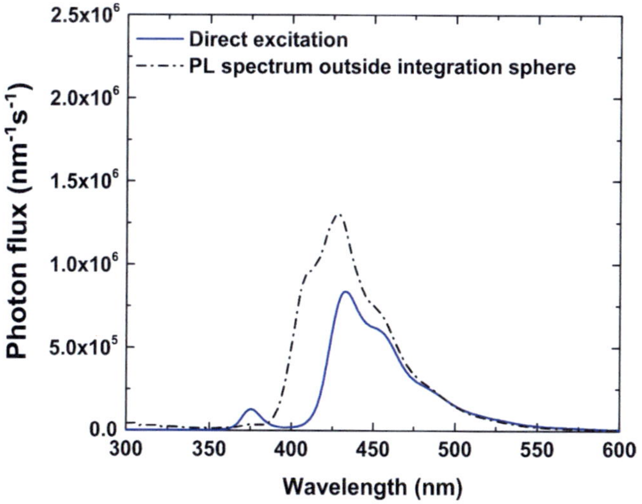

Figure 2.4: PL of Lumogen violet F570 BASF in a PMMA film that is recorded inside (solid) and outside (dashed) the integration sphere.

Reabsorption can be corrected by multiplying the spectrum outside the sphere so that the red part coincides with the spectrum recorded inside the sphere. The deviation of both spectrum is given by:

$$a = 1 - \frac{\int P_{inside}(\lambda)d\lambda}{\int P_{outside}(\lambda)d\lambda}, \tag{2.31}$$

where P_{inside} is the spectrum inside the sphere and $P_{outside}$ is the spectrum outside the sphere. By using the approach, that is considering reabsorption found by Wilson and Richards, the corrected PLQY is: [86]

$$\phi_{PLQY}^{corr} = \frac{\phi_{PLQY}}{1 - a + a\phi_{PLQY}}. \tag{2.32}$$

The derived formulas serve for the PLQY measurements that were performed in this thesis.

Excitation source and detectors

A 375 nm LED was used as an excitation source. The detector was a fiber-coupled Avantes spectrometer that was calibrated. The inner wall of the integration sphere was covered with PTFE (spectralon).

2.2.4 Streak camera for time-resolved PL measurements

A streak camera is a contraption that can resolve ultrafast light signals in time and space. The output is a two-dimensional matrix with color encoded intensity. Figure 2.5 shows an example of such a measurement. The scheme of the experimental setup is shown in Figure 2.6. The sample is pumped by a laser that comes from a femtosecond laser pumped optical parametric oscillator (OPO). After exciting the sample, the signal is collected by two collimation lenses, that focus the light into the spectrometer. The spectrometer guides the wavelength resolved signal into the streak camera. The streak camera is then performing the time resolution.

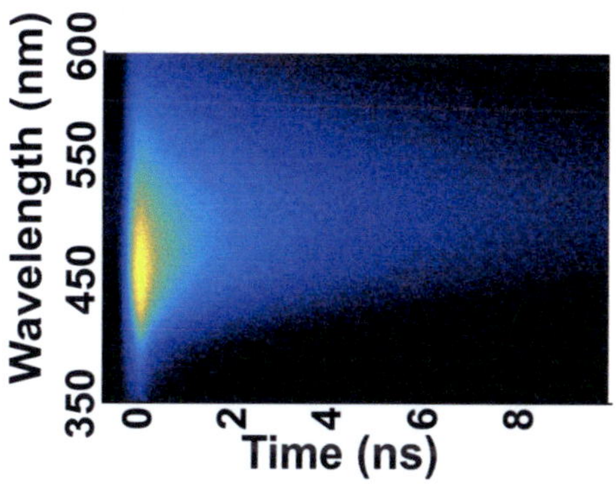

Figure 2.5: Example of a streak camera measurement. The color encoding is from black – low to blue – medium to yellow – high signal.

The delay generator is communicating the delay between the release of a pulse from the laser and the arriving of the emission in the streak camera. The complementary metal-oxide-semiconductor (CMOS) camera is transferring the 2D matrix onto the computer. The details of each part are explained below.

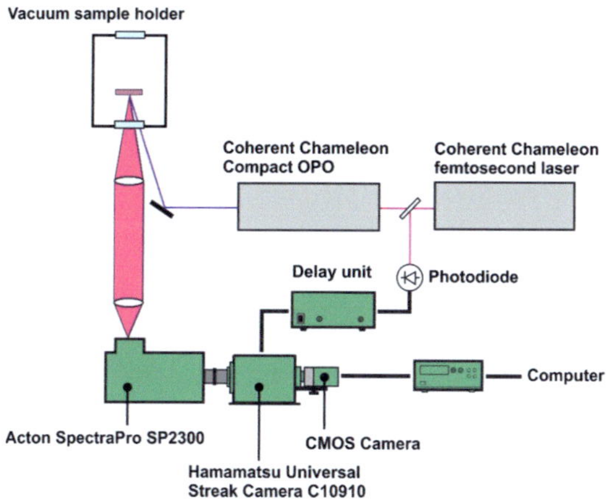

Figure 2.6: Scheme of the streak camera setup for time resolved PL measurements. Description in the main text. This figure is inspired by an experimental sketch provided by Hamamatsu's "Guide to Streak Cameras". [87]

Vacuum sample holder

The thin film solid-state samples were stored within an airtight vacuum sample holder to prevent them during the excitation from degradation (oxidation or reaction with water). The vacuum was created by using a turbo molecular vacuum pump and was kept at $<10^{-7}$ mBar during all measurements. If the samples were liquids, they cannot be put into a vacuum holder as they would start to boil at pressures lower than the vapor pressure of the used solvent at room temperature. Therefore, the solutions were prepared inside an argon flushed glovebox and encapsulated airtight.

Excitation source, Chameleon femtosecond laser and OPO

The excitation source was a Coherent Chameleon Compact OPO that was pumped by a mode locked Coherent Chameleon femtosecond laser. The repetition frequency of the femtosecond laser was 80 MHz corresponding to a time window between two pulses of 12.5 ns and a pulse width of ca. 100 fs. The OPO uses optical non-linear crystals that splits the pump laser frequency into an idler and a signal beam with frequencies which sum correspond to the pump beam:

$$\omega_P = \omega_I + \omega_S. \tag{2.33}$$

The pump, idler and signal beam are within an optical resonator and the non-linear crystal. The resonator is made from Kerr medium, which are non-linear mirrors that are transmissive at high light intensities but reflective at low light intensities. As soon as certain modes within the resonator reach the threshold of the Kerr medium they are released from the resonator. This is the output signal which is used for the excitation. This can be varied by temperature, orientation of the crystal or length of the resonator. [88]

Photodiode

The used photodiode is a pin-diode and receives the output from the femto-second pump laser by using a beam splitter that is located between the pump laser and the OPO in Figure 2.6. The pin-diode is a pn-diode with an intrinsic layer between the p and n doped regions. Due to the intrinsic region the depletion region is dissipating much slower than in a pn-diode making the pin-diode suitable for fast signals. That is needed, since the pump laser has a frequency of 80 MHz.

Delay unit, Stanford DG645

The delay unit is as the name says delaying an electrical signal by using a crystal-oscillator as an internal clock. After receiving an external signal, the delay generator starts to count the oscillations from the crystal-oscillator until the wished time delay is reached and gives then an electrical output trigger.

Spectrometer, Acton SpectraPro SP2300

The spectrometer uses a grating to spatially decompose the signal that comes from the sample after excitation. The spectrometer is resolving the emission signal spectrally.

Hamamatsu Universal Streak Camera C10910

The streak camera is the heart of the setup. After the spectrometer decomposed the signal spectrally the signal goes through a horizontal slit and two lenses that are focusing the light onto a photocathode. The photocathode releases electrons that fly through a vertical electrical field. The electrical field is deflecting the electrons onto a phosphor screen. The time resolution comes from the sweep of the deflecting electrical field.

CMOS Camera OrcaFlash 4.0 V2

The CMOS camera records the phosphor screen and transfers this to the computer. The idea behind a CMOS is that two transistors with complementary polarity are using the same ground material. If there is a voltage only the p-channel is conductive, corresponding to a logical 1 and if there is no voltage only the n-channel is conductive corresponding to a logical 0. The advantage of a CMOS in comparison to a transistor-transistor logic (TTL) are the lower operation voltage of around 0.75 V. In comparison, TTL used 5 V. The use of CMOS technology in a camera becomes obvious if one takes into account that neighboring pixels are interacting with each other. Lower voltages across each pixel is lowering these interactions.

2.2.5 High sensitivity setup for time-resolved and steady-state PL detection

Figure 2.7 shows the schematic of the high-sensitivity setup for PL measurements. For the steady-state measurements a Hamamatsu R928P photomultiplier tube (PMT) and a Stanford Research 850 lock-in amplifier are used. For the time-resolved measurements the Hamamatsu R298P PMT is used in photon counting mode (voltages above 1000 V) and a PicoQuant Time Harp 260 system. The setup uses a diode pumped solid state laser (DPSS) as an excitation source. The laser is filtered as sometimes side bands can appear that disturb the measurement. For the variation of the excitation power density a neutral density (ND) filter that is mounted onto a stepper motor is used. For the simultaneous measurement of a signal and the power a beam splitter (BS) is installed that reflects one part of the laser onto the sample and the other onto the power meter (PM). That is needed as the intensity of the DPSS laser is fluctuating over time. The laser is reflected by the BS onto the sample that is kept in a vacuum chamber or that was encapsulated in the glove box. The emission is collected by two collimation lenses (L) into the double monochromator. The double monochromator posses two grating that guarantee a very high spectral resolution. After passing through the double monochromator the signal reaches the photo multiplier tube (PMT). The PMT is then connected with the lock-in amplifier or the PicoQuant system. In the following the working principle of the significant parts of the experimental setup are explained.

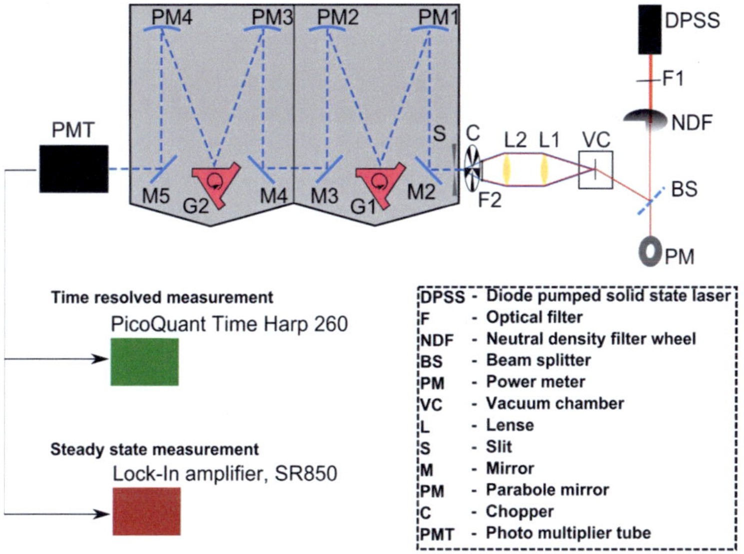

Figure 2.7: Scheme of high sensitivity PL detection setup. For time-resolved measurements the PicoQuant Time Harp 260 and for steady-state measurements the Stanford lock-in amplifier SR850 is used. This sketch was adapted from Kim's master thesis. [89]

DPSS laser source, Roithner RLTMSL-532

For most of the measurements a 532 nm DPSS continuous wave (cw) laser was used. For time resolved measurements the laser was electrically modulated. The common working principle of a 532 nm DPSS laser is using a powerful 808 nm pump laser coming from a GaAlAs electrically pumped laser diode. The laser diode is usually an electrically driven pin-diode. The intrinsic region increases the probability of recombination of the injected charge carriers. The generated light is surrounded by an optical cavity that forms a Fabry-Perot resonator. By doing so, the emitted light has a narrow wavelength distribution. Further, the cavity supports the amplified emission and the diode "lases". The generated pump laser is pumping a neodymium-doped yttrium orthovanadate crystal which produces 1064 nm laser emission. This emission is guided into a non-linear crystal ($KTiOPO_4$) that generates the second harmonic of 1024 nm, that is 532 nm.

Stepper motor

A stepper motor is essentially a normal electrical motor, that however is using several electrical magnets (stator) which are electrically driven and controlled. Instead of having an electrical drive with a constant frequency the electrical signal can have only one switching, which leads to the smallest possible rotation of the motor. The smallest rotation corresponds to the extension of one electrical magnet of the stator.

Photo multiplier tube, Hamamatsu R928P

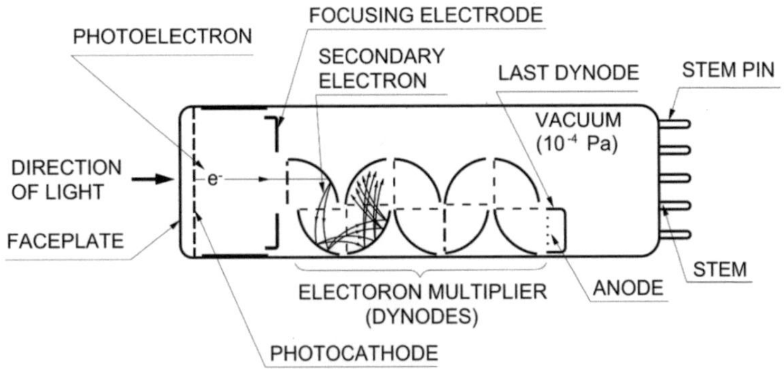

Figure 2.8: Scheme of a PMT. Adapted from Hamamatsu's manual. [90]

The photo multiplier tube (PMT) is the detector for the high sensitivity setup. Figure 2.8 shows a schematic of a PMT. The working principle is that light shines onto a metallic grid. Electrons are getting released and are focused on the first dynode plate. As a high voltage (~1000 V) is applied the electron accelerates onto the dynode and releases several electrons at its impact. The new generated electrons are accelerated onto the next dynode and release again new electrons. This process is repeated several times until from a single photon a huge number of electrons is accelerated towards the anode. This makes the PMT essentially a closed electrical circuit. It is necessary that the tube is evacuated so that the electrons are not disturbed or deflected by any other atoms. The current is then read out by an electrical amplifier that sends the signal to the lock-in amplifier or the PicoQuant.

Steady-state measurement, Lock-in amplifier Stanford SR830 DSP

This setup was used for the recording of weak emission spectra and for excitation power dependent measurements, that must cover a large range (3-4 orders) of excitation power densities. Especially at lower excitation power densities, the setup played a significant role. The lock-in amplifier played here an important role.

The centerpiece of the steady-state measurement is the lock-in amplifier. The principle of a lock-in amplifier is that the signal is time modulated by a chopper. The chopper signal is then multiplied with the input signal from the PMT and integrated over time:

$$U(t) = \frac{1}{T}\int_{t-T}^{T} U_{ref}\left(\frac{\tau}{T} + \Delta\phi\right) U_{in}\left(\frac{\tau}{T}\right) d\tau. \tag{2.34}$$

The input and the reference signal are time modulated with a fixed period T and corresponding frequency $\frac{2\pi}{T} = \omega$ and with a phase difference $\Delta\phi$. As any periodic signal can be expressed by a sinus function (Fourier transform).

Hence, following integral is representing the generic case:

$$U(t) \propto \int_{t-T}^{T} \sin(\omega_{ref}\tau + \Delta\phi)\sin(\omega_{in}\tau)\,d\tau \tag{2.35}$$

Integrating this over time would result in zero, except for the case when $\omega_{ref} = \omega_{in}$. In that case the outcome is a DC signal:

$$U(t) \propto \cos(\Delta\phi). \tag{2.36}$$

This means that only a signal that has the same frequency as the chopper is measured and can be further amplified. This reduces significantly noise, that has a rather random distribution of frequencies. Care has to be taken of reflections, that will have another phase and can get amplified instead of the actual signal. Further, the laser signal itself has to be properly blocked (via optical filter) before entering the double monochromator. For very low intensities, the chopping frequency should be very high and at the same time the integration time should be very high as well. This guarantees that the lock-in amplifier is integrating over as many periods as possible resulting in a cleaner signal output. This however increases the measuring time significantly.

Time-resolved measurements, PicoQuant TimeHarp 260

The time-resolved measurements have been performed by using the method of time-correlated singe photon counting. The working principle is that a trigger signal (electrical trigger from the laser time modulation) starts an internal clock, that starts to run until a second signal (single photon from the excited sample) releases an electrical signal that turns off the clock. This is repeated several times until a decay curve is recorded.

2.3 Synthesis of surface-anchored metal-organic frameworks

The synthesis of SURMOFs is rather simple as it does not require any special techniques. In the following, two methods are presented. The first is the spray casting method and is historically the first verified method to grow SURMOFs. In comparison to this the spin coating method is the second procedure to grow SURMOFs. The advantage of spin coating is that it produces on a nanoscale smoother films, however, the growth per cycle is significantly lower. That is not a disadvantage, as spin coating gives higher thickness control compared to spray casting. However, the spin coating method was developed only in the last several months and was only used for the latest SURMOF samples.

Spray casting method

For the synthesis of SURMOFs by spray casting method a linker solution, a zinc acetate solution and a rinsing solution have to be prepared. The SURMOFs prepared in this work are based only on ethanol as a solvent and as a rinsing solution. Following experience values for the concentrations have been chosen to $c_{linker} = 20\mu M/L$ and for zinc actetate $c_{zinc\ acetate} = 0.2mM/L$. [39]

The preferred substrates have a native oxide layer; these are either glass slides (for spectroscopic analysis) or silicon wafers (for structural analysis). The chosen substrate must be cleaned thoroughly in a 1:1 acetone ethanol solution in an ultrasonic bath for ~5 minutes. After cleaning, the native oxide layer can be functionalized by inducing hydroxylic groups by either oxygen plasma or ultraviolet light currying (in each case for 15 minutes). Significant differences in the quality of the prepared films with respect to the functionalization procedure are not observed. Therefore, ultraviolet light curing is preferred.

After cleaning and functionalization, the zinc acetate solution is sprayed onto the substrate for 15 seconds, followed by a waiting time of 35 seconds. Afterwards, the rinsing solution is used for 5 seconds to remove not-reacted material. After rinsing, the linker solution is sprayed onto the substrate for 25 seconds. After a waiting time of 35 seconds the residuals are washed away by the rinsing solution for 5 seconds and the cycle is repeated until the desired thickness of the film is reached.

During the waiting-steps, substrates are tilted horizontally as it was found that this leads to optically highly qualitative films. Further, a value for the thickness growth is estimated in the case of spray casting being around ~7 times the lattice constant (that depends on the used organic linker) in z-direction of the grown system. For 4,4'-(anthracene-9,10-diyl)dibenzoate based SURMOF (lattice constant 2 nm) it is found to be 15 nm per cycle and for Pd(II) 5,15-diphenyl-10,20-di(4-carboxyphenyl) porphyrin based SURMOF (lattice constant 2.4 nm) a growth rate of roughly 16 nm per cycle is found.

Spin coating method

The preparation of SURMOFs by spin coating follows the same prerequisites as for spray casting – three solutions and a functionalized substrate. After the preparation of the substrate, it is positioned in a spin coater. 100µL of metal solution is drop casted onto the substrate that is rotated at 5 kHz for 30 seconds. Afterwards, ethanol is drop casted (100µL) onto the substrate to remove non-reacted residuals, followed by the dropcast of the linker solution (100µL) and a subsequent drop cast of ethanol (100µL). SEM images have shown that the growth rate per cycle is close to one times the lattice constant.

2.4 Structure analysis of surface-anchored metal-organic frameworks

X-ray diffractometry

SURMOFs are organic crystalline structures and so capable for X-ray diffractometry (XRD) characterization. The main concept is that X-rays that are reflected from two parallel crystal surfaces are interfering. For constructive interference Bragg's law applies:

$$n\lambda = 2d\sin(\theta), \tag{2.37}$$

here n is any integer representing the order of constructive interference, λ is the wavelength of used x-rays, d the space between neighbouring crystal surfaces and θ is the incident angle. The X-ray diffractograms (XRD) are recorded using a Bruker D8 Advance. The machine uses a Cu-anode with a Cu $\mathrm{K}_\alpha 1{,}2$ radiation with a wavelength of λ=0.15419 nm. As SURMOFs are grown as thin films on planar substrates there are two configurations on how to measure the XRD. The first configuration is out-of-plane that is used to characterize the crystallinity with respect to the substrate. The second configuration is in-plane and is used to characterize the crystallinity parallel to the substrate. Figure 2.9 illustrates both configurations.

XRD was used to verify, that only a single orientation of the SURMOF is presented. Additionally, XRD could be used to identify whether multilayer SURMOFs can be formed. The resulting XRD of a multilayer SURMOF is a sum of the individual SURMOF XRDs.

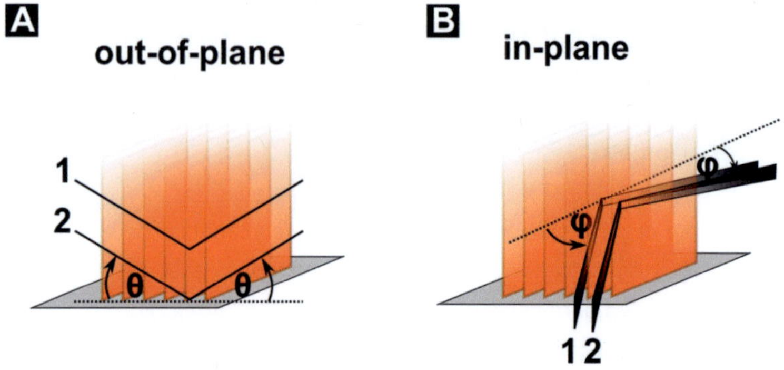

Figure 2.9: Schematic of a SURMOF (orange) on a substrate (grey). A) The path of beam 1 and 2 is shown for the out-of-plane configuration. θ is the incident angle. B) The path of beam 1 and 2 is shown for the in-plane configuration. For clarity the beams are shown as 2 dimensional planes that are parallel to the substrate. The incident angle is φ.

Time-of-flight secondary ion mass spectroscopy

For the multilayer analysis Time-of-flight secondary ion mass spectroscopy measurement was performed for cross sectional material composition analysis. Bi^{3+} ions are used to erode the material. As the ion fluence increases with time on the x-axis, the composition of the sample through its depth is roughly revealed. This method, however, is limited as SURMOFs are soft materials, meaning that the impact of the Bi^{3+} ions will go deep 10-50 nm into the SURMOF.

However, using both characterization techniques (XRD and ToF-SIMS) provided enough evidence that in a multilayer SURMOF each SURMOF is intact.

Infrared spectroscopy

XRD is used to get the orientation and structure of the SURMOF, however, non-reacted compounds cannot be detected by XRD. Infrared spectroscopy was used to characterize whether all linkers have reacted and formed the paddle-wheel formation or how infiltrated dye molecules have reacted or bonded within the SURMOF. Therefore, infrared spectroscopy is used. By shining infrared light onto the SURMOF certain frequency bands are absorbed by the SURMOF. These frequencies correspond to molecular vibrations within the SURMOF. For the characterization of SURMOFs that are ideally grown on gold substrates the technique of Infrared reflection-absorption spectroscopy (IRRAS) is used. For this an infrared beam is directed under the grazing incident angle (ca. 88°) onto the thin film. By doing so, the SURMOF is absorbing certain frequencies of the IR beam and that can be seen in the reflected beam. The IRRA spectra are acquired with a resolution of 2 cm^{-1} using a FTIR spectrometer (Bruker VERTEX 80) under ambient conditions, with an angle of incidence of 80° relative to the surface normal and a liquid nitrogen cooled mercury cadmium telluride (MCT) narrow band detector.

Scanning electron microscopy

For the morphological studies (top and side view) of the created SURMOFS scanning electron microscopy (SEM) is used. The working principle of the SEM are accelerated electrons, that are focused by electromagnetic lenses onto the area of interest. The impact of electrons results in various interactions with the media like X-rays, visible light, scattered electrons or secondary electrons. The later are used for the SEM image.

This helps to obtain information on continuity, compactness, and homogeneity of the SURMOF thin films. For cross-sectional SEM images a LEO 1530 Gemini scanning microscope at 3–5 kV is used. However, at high electron acceleration voltages the SURMOF started to degrade. Therefore, the scanning voltage was limited to 3 kV.

3 Theory of Organic Spectral Upconversion

In this chapter the theoretical foundations are described for triplet-triplet annihilation based upconversion. This chapter has three big parts. The first part describes the dynamics of excited states by using rate equations. In the second part these are verified by experimental data. Based on the experimentally verified theory a new hypothesis is derived for an emitter sensitizer bilayer system in the third part.

3.1 Rate equation formulation for isotropic systems

The time dynamics of TTA-UC can be elegantly formulated mathematically by using the rate equation approach. The underlying assumption of rate equations is a continuous distribution of the excited states in space. The comparison between theoretical solutions, whether numerical or analytical, and data obtained from experiments can lead to a better understanding of the underlying mechanisms. The analysis starts with the absorption of a photon by a sensitizer molecule. The excited sensitizer singlet state can relaxate very quickly by fluorescence to its ground state, which would make an energy transfer onto an emitter molecule inefficient. Therefore, sensitizer molecules in efficient upconversion systems have to perform fast ISC. ISC flips the spin of the excited electron due to spin-orbit coupling (induced by a heavy atom); the singlet exciton transforms into a triplet exciton. The following rate equation describes the singlet exciton population on the sensitizer molecule:

$$\frac{\partial S_S}{\partial t} = G\big(S_0, I(x)\big) - k_S^{\mathrm{fl}} S_S - k_{\mathrm{ISC}} S_S. \tag{3.1}$$

Here, S_S describes the concentration of sensitizer molecules in the singlet state, k_S^{fl} describes the monomolecular decay rate via fluorescence, k_{ISC} the ISC rate, and $G(S_0, I(x))$ is the generation rate of sensitizer singlet excitons. The generation rate depends on the concentration of sensitizer molecules in the ground state S_0. Furthermore, it is a function of the excitation power density of the illumination source that is a function of the photon flux N:

$$I(x) = N(x)\frac{\mathrm{hc}}{\lambda}. \tag{3.2}$$

where λ is the wavelength, h is the Planck constant and c is the speed of light. In the following, it is assumed that the concentration of sensitizer molecules in the ground state is huge compared to any concentration of excited sensitizer molecules. Within this approximation, the absorption of photons is intensity independent. The generation rate according to Lambert-Beer's law:

$$G(x) = \alpha I_0 \exp(-\alpha d), \tag{3.3}$$

here α is the absorption coefficient, I_0 is the illumination power density and d is the beam path through the system. For weak absorption $\alpha^{-1} \gg d$ equation (3.3) simplifies to:

$$G = \alpha I_0. \tag{3.4}$$

After ISC the created triplet excitons can transfer onto an emitter molecule. The likelihood for energy transfer is high because the lifetime of triplet states is long compared to the lifetime of singlet states. The process of triplet states from the sensitizer to the emitter is DET. With this the concentration of the triplet excited states on the sensitizer molecule can be described as follows:

$$\frac{\partial S_T}{\partial t} = k_{\mathrm{ISC}} S_S - k_S^D S_T - k_S^{\mathrm{DET}} S_T. \tag{3.5}$$

Here, k_S^D is the monomolecular decay rate of sensitizer triplets and k_S^{DET} the energy transfer rate by DET. The sensitizer triplets are generated from the sensitizer singlets by ISC. If the ISC rate k_{ISC} is much larger than any other monomolecular decay rate of the sensitizer, the concentration of the sensitizer singlet state reaches steady-state fast so that equation (3.1) can be set to zero and solved for S_S which is inserted into equation (3.5). By introducing the ISC efficiency $\phi_{\mathrm{ISC}} = k_{\mathrm{ISC}}/\left(k_{\mathrm{ISC}} + k_S^{\mathrm{fl}}\right)$, equation (3.5) simplifies to:

$$\frac{\partial S_T}{\partial t} = \phi_{\mathrm{ISC}} G - k_S^D S_T - k_S^{\mathrm{DET}} S_T. \tag{3.6}$$

After sensitization of the emitter triplet level, the emitter triplet decays either non-radiatively or via TTA provided that a collision with another emitter triplet occurred. With this, the time dynamics of the emitter triplet excitons are described by:

$$\frac{\partial E_T}{\partial t} = k_S^{\mathrm{DET}} S_T - k_E^D E_T - \gamma E_T^2, \tag{3.7}$$

where k_E^D is the monomolecular decay rate of emitter triplets and γ the collision frequency of TTA of two emitter triplets. The singlet state of the emitter is ideally populated by the TTA process and decays by the emission of a high energy photon. Finally, the time dynamics of the emitter singlets are:

$$\frac{\partial E_S}{\partial t} = \frac{1}{2} f \gamma E_T^2 - k_E^{fl} E_S, \tag{3.8}$$

where the factor of $1/2$ means that after the collision of two emitter triplets there is only one emitter singlet. f is a spin statistics factor as two triplets can result in one singlet, three triplets and five quintets can result, we assume $f = 1$ (only singlets get generated), and k_E^{fl} is the monomolecular fluorescence decay rate. Following the decay channels from generation to the release of a high energy photon the upconversion efficiency ϕ_{UC} is described by:

$$\phi_{\mathrm{UC}} = \frac{k_E^{\mathrm{fl}} E_S}{G} = \frac{1}{2} f \phi_{\mathrm{ISC}} \phi_{\mathrm{DET}} \phi_{\mathrm{TTA}} \phi_{\mathrm{PLQY}}, \tag{3.9}$$

$\phi_{\mathrm{DET}} = k_S^{\mathrm{DET}}/(k_S^{\mathrm{DET}} + k_S^D)$ is the DET efficiency, $\phi_{\mathrm{TTA}} = \gamma E_T^2/(\gamma E_T^2 + k_E^D)$ is the TTA efficiency, and ϕ_{PLQY} is the emission quantum yield of the emitter molecule. Equation (3.6) to (3.9) are the basic equations describing organic upconversion and are now further analyzed.

3.1.1 Steady-state solution

The described equations cannot be solved fully analytically. However, certain approximations can be made leading towards analytic solutions that improve the understanding of the upconversion system. The first approximation is that the system is excited by a continuous excitation source and is therefore driven into steady-state meaning that there is no change in concentrations over time:

$$0 = G - k_S^D S_T - k_S^{\mathrm{DET}} S_T \tag{3.10}$$

$$0 = k_S^{\mathrm{DET}} S_T - k_E^D E_T - \gamma E_T^2, \tag{3.11}$$

$$0 = \frac{1}{2} \gamma E_T^2 - k_E^{\mathrm{fl}} E_S, \tag{3.12}$$

Here, $\phi_{\mathrm{ISC}} = 1$ is assumed. Solving equations (3.10) to (3.12) results in the excitation intensity dependence of the upconversion intensity:

$$I_{\mathrm{UC}} = k_E^{\mathrm{fl}} E_S = \frac{1}{8} \frac{k_E^{D^2}}{\gamma} \left(\sqrt{1 + 4 \frac{G}{G_{TH}}} - 1 \right)^2. \tag{3.13}$$

Here $\frac{1}{\phi_{DET}}\frac{k_E^{D^2}}{\gamma}$ is the upconversion threshold G_{TH}, that is a characteristic generation rate and a figure of merit of TTA-UC. For the beauty of simplicity equation (3.13) is normalized by $8\,\gamma/k_E^{D^2}$ and is transformed into a dimensionless form (a tilde will mark in the following dimensionless quantities):

$$\tilde{I}_{\mathrm{UC}} = \left(\sqrt{1 + 4\frac{G}{G_{\mathrm{TH}}}} - 1\right)^2. \tag{3.14}$$

The so derived intensity dependence has two limits concerning the upconversion threshold G_{TH}. The first limit is at generation rates much smaller than the upconversion threshold: $G \ll G_{\mathrm{TH}}$,

$$\tilde{I}_{\mathrm{UC}}^{G\ll G_{\mathrm{TH}}} = 4\frac{G^2}{G_{\mathrm{TH}}^2}. \tag{3.15}$$

In this limit, the upconversion emission intensity is proportional to the square of the generation rate. From equation (3.9) follows that then the upconversion efficiency is linearly dependent on the generation rate. The other limit is for generation rates much higher than the upconversion threshold $G \gg G_{\mathrm{TH}}$:

$$\tilde{I}_{\mathrm{UC}}^{G\gg G_{\mathrm{TH}}} = 4\frac{G}{G_{\mathrm{TH}}}. \tag{3.16}$$

In this limit, the upconversion emission intensity depends linearly on the excitation power density and the upconversion efficiency is constant. The intersection point of the two described limits is at $G = G_{\mathrm{TH}}$. With that, the upconversion threshold G_{TH} sets the excitation power density from which the upconversion efficiency approaches the maximum efficiency.

$$G_{\mathrm{TH}} = \alpha I_{\mathrm{TH}} = \frac{1}{\phi_{\mathrm{DET}}} \frac{{k_E^D}^2}{\gamma}. \tag{3.17}$$

Equation (3.17) shows which system parameter has to be optimized for achieving a low I_{TH}. The sensitizer molecule must fulfill the following requirements:

1) the absorption coefficient α has to be as large as possible, meaning that the concentration generated sensitizer triplets is maximized,

2) ϕ_{DET} must be close to unity so that all sensitizer triplets are transferred onto the emitter molecules.

The emitter molecule has to fulfill the following requirements:

1) the monomolecular decay rate k_E^D must be small, or in other words, the emitter triplet needs to have a long lifetime,

2) the collision frequency γ must be high.

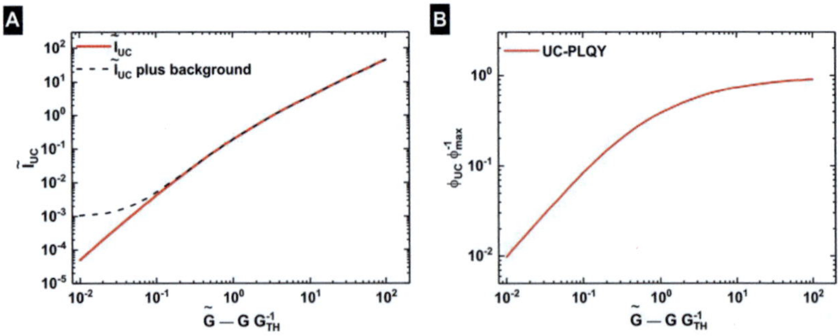

Figure 3.1: A) The upconversion intensity dependence (solid curve) is shown as a function of $\tilde{G} = G\, G_{TH}^{-1}$. B) The intensity dependence for the quantum yield (solid curve) is plotted as a function of $\tilde{G} = G\, G_{TH}^{-1}$.

In other words, the bimolecular decay rate γE_T should dominate the monomolecular decay rate k_E^D for a low threshold. Hence, the threshold also defines the transition point between a predominantly monomolecular to a predominantly bimolecular emitter triplet decay. This can be shown by solving equations (3.10) to (3.12) by neglecting the bimolecular decay rate, which leads to a quadratic intensity dependence of the upconversion emission:

$$I_{\mathrm{UC}} = \frac{1}{2}\gamma \frac{\phi_{\mathrm{DET}}^2}{k_E^{D^2}} G^2. \tag{3.18}$$

Neglecting the monomolecular decay rate results in a linear dependence of the excitation power intensity:

$$I_{\mathrm{UC}} = \frac{1}{2}\phi_{\mathrm{DET}} G. \tag{3.19}$$

The intersection of both asymptotes again gives the upconversion threshold. Figure 3.1A illustrates the derived intensity dependence.

The derived steady-state behavior and especially the upconversion threshold give a first experimental method for the analysis of a TTA-UC system. However, the threshold is a constant that is a combination of all relevant parameters. Meaning that the measurement of the intensity dependence does not give any detailed information on these parameters. A method to extract the values for the emitter triplet decay rate k_E^D, the DET efficiency ϕ_{DET} and the collision frequency γ is to analyze the time-dependent solutions of the provided rate equations.

3.1.2 Simplified time-dependent rate equations

Equations (3.6) to (3.8) are coupled systems of non-linear first order differential equations. The non-linearity of the equations does not allow full solutions in the form of standard algebraic functions. However, assumptions can be made that lead to simpler equations and the corresponding solutions are describing the system good enough. The transformation of equations (3.6) to (3.8) into a dimensionless form will help in making certain assumptions obvious. For this, all three equations are multiplied by $\gamma/{k_E^D}^2$ and the following transformations performed:

$$\frac{\gamma X}{k_E^D} \to \tilde{X}, k_E^D t \to \tau_E, \frac{G}{G_{\mathrm{TH}}} \to \tilde{G} \tag{3.20}$$

where X stands for any concentration and t is the time. In the following $\tilde{X}$ is referred to as concentration, τ_E as time and $\tilde{G}$ as generation rate although these have not the same meaning as the physical quantities. The transformed dimensionless rate equations are:

$$\frac{\partial \tilde{S}_T}{\partial \tau_E} = \frac{1}{\phi_{\mathrm{DET}}}\tilde{G} - \frac{k_S^D + k_S^{\mathrm{DET}}}{k_E^D}\tilde{S}_T, \tag{3.21}$$

$$\frac{\partial \tilde{E}_T}{\partial \tau_E} = \frac{k_S^{\mathrm{DET}}}{k_E^D}\tilde{S}_T - \left(1 - \tilde{E}_T\right)\tilde{E}_T, \tag{3.22}$$

$$\frac{\partial \tilde{E}_S}{\partial \tau_E} = \frac{1}{2}\tilde{E}_T^2 - \frac{k_E^{fl}}{k_E^D}\tilde{E}_S. \tag{3.23}$$

In an efficient upconversion system with a low threshold, the sensitizer triplets are transferred faster to the emitter than the emitter triplets are decaying. Formally this means that $\frac{k_S^D + k_S^{\mathrm{DET}}}{k_E^D} \gg 1$. Finally, this means that the sensitizer triplet concentration reaches steady-state much faster than the emitter triplet concentration. Furthermore, the monomolecular decay rate of the emitter singlets, that is the fluorescence decay rate is much quicker than the monomolecular decay rate of the emitter triplets, $k_E^{fl}/k_E^D \gg 1$. Both assumptions mean that $\tilde{S}_T$ and $\tilde{E}_S$ reach steady-state faster than $\tilde{E}_T$. Ultimately this means that in this limit only the emitter triplet concentration is time-dependent. With this assumption, the three differential equations reduce to one single differential equation describing $\tilde{E}_T$ and one algebraic equation for $\tilde{E}_S$:

$$\frac{\partial \tilde{E}_T}{\partial \tau_E} = \tilde{G} - \left(1 - \tilde{E}_T\right)\tilde{E}_T, \tag{3.24}$$

$$\frac{k_E^{fl}}{k_E^D}\tilde{E}_S = \frac{1}{2}\tilde{E}_T^2. \tag{3.25}$$

3.1.3 Rising edge under quasi cw excitation, steady-state time

For a time-resolved analysis the excitation source should be time modulated. Here, the case for a quasi cw excitation is analyzed. Quasi cw means here that the excitation source is square wave modulated, with an on-time T_0. Ideally, the on-time of the excitation is longer than the system needs to approach steady-state. However, it is not clear from the equations when steady-state is reached. For an estimate of the steady-state time, the time-wise rising of the emitter triplets is analyzed. By using the boundary condition of no emitter triplets $\tilde{E}(\tau_E = 0) = 0$ and a constant generation rate a steady-state time for the emitter triplets and consequently for the emitter singlets can be found. Equation (3.24) is of Riccati type and can be solved analytically by using the defined boundary conditions:

$$\tilde{E}_T = \frac{2\tilde{G}}{1 + \sqrt{1 + 4\tilde{G}}\coth\left(\sqrt{1 + 4\tilde{G}}\frac{\tau_E}{2}\right)}. \tag{3.26}$$

The concentration saturates for $\tau_E \to \infty$ corresponding to the steady-state solution $\tilde{E}_T(\tau_E \to \infty) = \tilde{E}_T^{SS} = \frac{1}{2}\left(1 + \sqrt{1 - 4\tilde{G}}\right)$. For $\tau_E \to 0$ the triplet concentration rises linearly in time $\tilde{E}_T = \tilde{G}\tau_E$. The steady-state time can be defined by the intersection of the linear increase and the saturation value:

$$\tau_E^{SS} = \frac{1}{2\tilde{G}}\left(\sqrt{1+4\tilde{G}}-1\right). \tag{3.27}$$

As expected τ_E^{SS} is zero when the excitation power density is infinite and approaches the emitter triplet lifetime $\tau_E^{SS} = 1$ in the limit of small excitation power density. Equation (3.27) is underestimating the real steady-state time of the emitter triplets slightly, due to the assumption made that the sensitizer triplets are reaching steady-state instantaneously. This can be corrected by:

$$\Delta\tau_E^{SS} = \frac{k_E^D}{k_S^D + k_S^{DET}}. \tag{3.28}$$

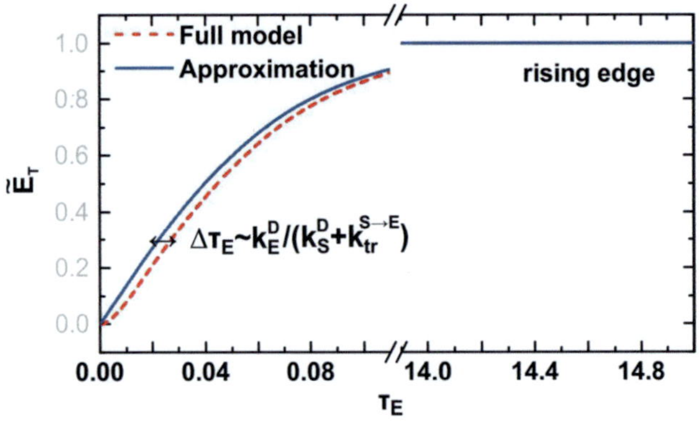

Figure 3.2: The rising edge of the emitter triplet concentration $\tilde{E}_T$ as a function of time for the particular case $\tilde{G} = 1$. Both, the approximated curve derived in the main text (solid curve) and the numerically calculated (dashed curve) are shown. Inset is the difference between the approximated and the analytical solution.

Figure 3.2 shows the deviation between the full model and the derived approximated solution. The difference is indicated by a horizontal double-arrow and the correction factor $\Delta\tau_E$. An estimation of the steady-state time of the emitter singlet excitons $\tilde{E}_S$ that corresponds to the observable PL UC emission equation (3.25) can be used together with equation (3.26).

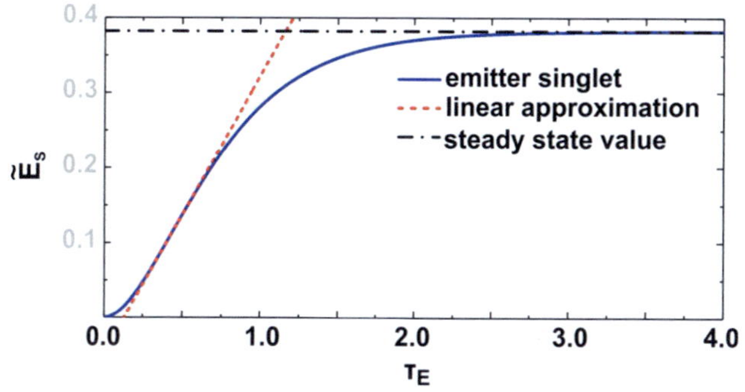

Figure 3.3: The emitter singlet concentration (solid curve) is plotted as a function of time for the particular case of $\tilde{G} = 1$. The linear approximation (dashed curve) around the inflection point is plotted together with the steady-state value (dashed-dotted curve).

Figure 3.3 illustrates the time-dynamics of $\tilde{E}_S$ for an arbitrary generation rate. A linear approximation around the origin leads to a constant value. Therefore for a proper estimation of the steady-state time of the emitter singlets, a linear approximation around the inflection point has to be performed. The intersection of the linear approximation and the saturation value defines the experimentally observable steady-state time. Figure 3.3 illustrates this approach. As for $\tilde{E}_T$ the steady-state time for the emitter singlets $\tilde{E}_S$ reaches a maximum for $\tilde{G} \to 0$. Hence, the easiest approach is to solve equation (3.24) in the monomolecular approximation by neglecting the quadratic term in $\tilde{E}_T$ and inserting this solution into equation (3.25). The solution, , in this case,, is given by:

$$\tilde{E}_S^{MM} = \frac{1}{2} \tilde{E}_T^{MM^2} = \frac{\tilde{G}^2}{2} (1 - \exp(-\tau_E))^2 \tag{3.29}$$

the inflection point of this curve is at $\tau_{\text{inflection}} = \ln(2)$ and the corresponding linear approximation at $\tau_{\text{inflection}}$ is:

$$\tilde{E}_S^{\mathrm{MM}}(\tau_E = \tau_{\mathrm{inflection}}) \approx \frac{1}{4}\tilde{G}^2\left(\frac{1}{2} + (\tau - \ln(2))\right). \tag{3.30}$$

The intersection of the linear approximation and the steady-state value $\tilde{G}^2/2$ gives the steady-state time in the limit $\tilde{G} \to 0$:

$$\tau_{\mathrm{UC}}^{\mathrm{SS}} = \frac{1}{2}(3 + 2\ln(2)) \approx 2.2. \tag{3.31}$$

The found steady-state time is roughly double the lifetime of an emitter triplet. In other words, at least two emitter triplet excitons have to be excited until TTA happens. This estimation of a maximum steady-state time for upconversion is important for a time-dependent analysis. An excitation pulse has to be at least twice as long as the emitter triplet lifetime until steady-state is reached.

3.1.4 Decay after quasi continuous wave excitation

The decay of the emission after excitation is analyzed by setting the boundary condition of an emitter triplet concentration at steady-state $\tilde{E}_T(\tau_E = 0) = \tilde{E}_T^{SS}$ and turned off generation rate $\tilde{G} = 0$. With this equation (3.24) is reduced to:

$$\frac{\partial \tilde{E}_T}{\partial \tau_E} = (1 - \tilde{E}_T)\tilde{E}_T, \tag{3.32}$$

and the solution is:

$$\tilde{E}_T = \frac{1}{\frac{1}{\phi_{\mathrm{TTA}}}\exp(\tau_E) - 1}. \tag{3.33}$$

In the limit of $\tilde{G} \to 0$ TTA efficiency at steady-state drops to zero and the emitter triplet concentration decays purely monomolecular:

$$\tilde{E}_T^M = \phi_{\mathrm{TTA}} \exp(-\tau_E) \tag{3.34}$$

This helps to extract the emitter triplet lifetime from the upconversion emission, as the upconversion emission in the monomolecular approximation is (inserting equation (3.34) into (3.25)):

$$\tilde{E}_S^M = \frac{1}{2} \phi_{\mathrm{TTA}}^2 \exp(-2\tau_E) \tag{3.35}$$

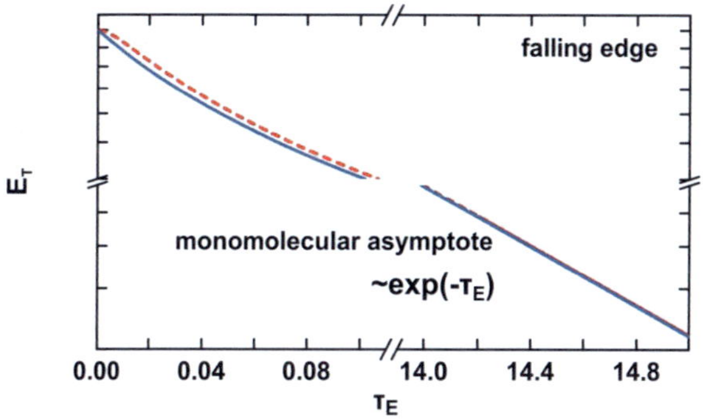

Figure 3.4: The emitter triplet concentration is shown for the falling edge. Both, the approximated solution (solid curve) and the exact solution (dashed line) are approaching the monomolecular limit inset as an equation. The

Figure 3.4 shows the plot for the derived solution, as well as the numerically found solution. Again, there is a slight deviation between the full rate equation model and the found solution. And yet this is due to the approximation made for the sensitizer triplet concentration reaching steady-state instantaneously.

3.1.5 Summary of time-dependent analysis

Figure 3.5 shows the normalized plots of the derived rising and the falling edge for $\tilde{E}_S$ at different excitation power densities, spanning six orders of magnitude. Additionally, the monomolecular and bimolecular asymptotes are plotted by dashed curves marking the two limits for a time-dependent analysis.

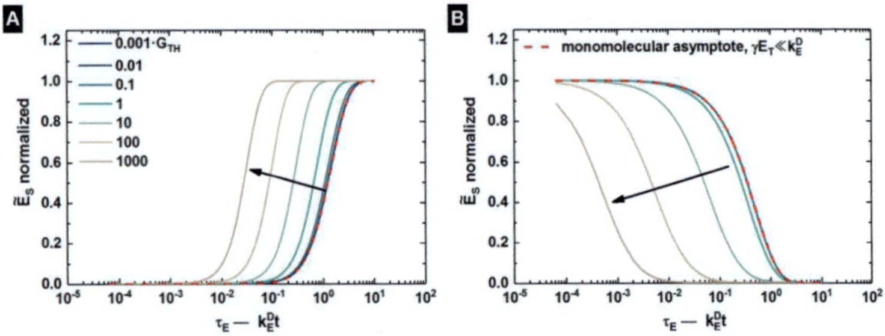

Figure 3.5: Shown are the rising and falling edge of $\tilde{E}_S$ as a function of τ_E at different excitation power densities (solid curves). Additionally, the expected curves for the bimolecular (dotted curves) and monomolecular (dashed curves) asymptotes are shown. The arrow indicates the direction of rising excitation power density.

The provided analysis of the time-dependent solutions gives a framework for the experimental measurements. From the falling edge the decay parameters γ and k_E^D can be extracted. However, to provide proper initial value conditions, the system has to be driven into the steady-state. The minimum duration of a laser pulse has to be at least twice as large as the emitter triplet lifetime to guarantee that the steady-state concentrations given by the rate equations are reached. In the next section, a model system is analyzed for the verification of the provided theoretical analysis.

3.2 Comparison between theory and experimental data

The derived rate equations are strongly valid for systems in which the emitter and the sensitizer molecules are uniformly distributed. Therefore, a solution-based system is the best choice to verify that theory and experiment match.

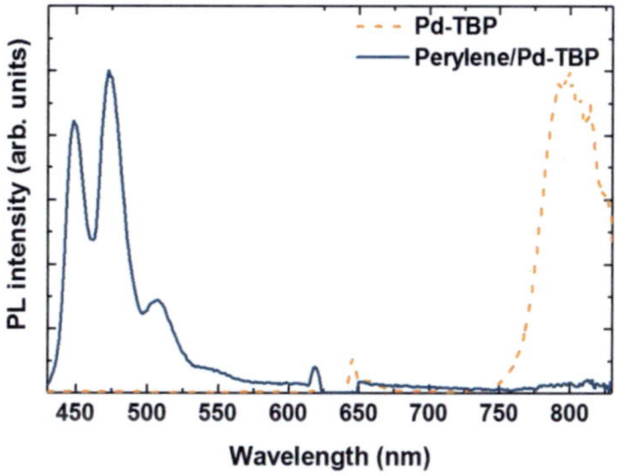

Figure 3.6: Photoluminescence spectra are shown for a Pd-tetrabenzoporphyrin (Pd-TBP) solution (dashed curve) and the mixed solution perylene/Pd-TBP (solid curve).

Figure 3.6 shows the photoluminescence spectrum for a perylene/Pd-tetrabenzoporpyhrin (Pd-TBP) system in toluene solution excited with a laser diode at 633 nm. The 1 mm thick cuvette is filled with the solution in a glove box and closed airtight to prevent triplet quenching by oxygen. From power measurements, the fraction of photons absorbed is estimated to be 0.44, which results in α =4.4 cm^{-1}. The PL spectrum in Figure 3.6 shows that in the presence of perylene the phosphorescence signal at 800 nm is strongly quenched indicating that ϕ_{tr} is close to unity and strong upconversion emission at ~ 473 nm is present.

Steady-state analysis

The intensity dependent measurement is shown in Figure 3.7. A modified version of equation (3.13) is used to estimate the threshold of this system:

$$I_{\mathrm{UC}} = A \cdot \left(\sqrt{1 + 4 \frac{G}{G_{TH}}} - 1 \right)^2 + C. \tag{3.36}$$

Here, A is an arbitrary amplitude and C is a background level. From the fit, a threshold $\mathrm{I_{TH}} = (11.3 \pm 0.6)\,\mathrm{mW\,cm^{-2}}$ corresponding to a generation rate

$G_{TH} = \alpha I_{TH} = (3.6 \pm 0.2) \cdot 10^{16} s^{-1} cm^{-3}$ is estimated. The relative error between data and fit is shown in the bottom panel of Figure 3.7 and only below $1\,mW\,cm^{-2}$, the error is significantly higher than 10%. Further, the intensity dependence shows no bending of the curve below a slope of 1 at high excitation power densities. This behaviour proves that non-linear behavior of the sensitizer is not present at the chosen concentrations, as is demonstrated in similar systems.[13]

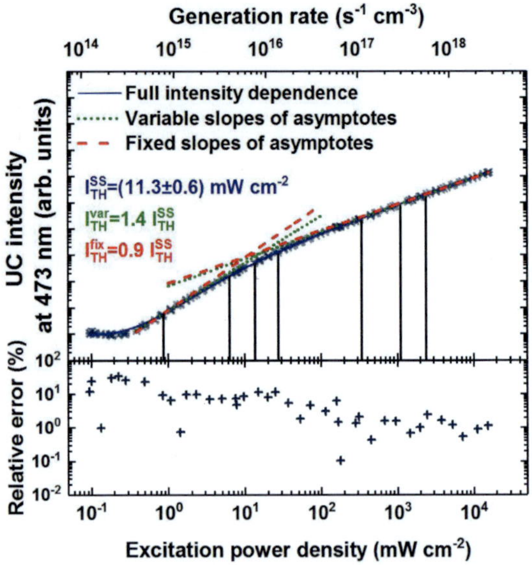

Figure 3.7: In the top panel, the intensity dependence for the upconversion emission at 473nm is shown (crosses). For clarity, every third point is hidden. The intensity dependence is fitted to the steady-state intensity dependence (solid line). From the fit, an upconversion threshold I_{TH}^{SS} is extracted. Additionally, the two limiting asymptotes with a variable (dotted lines) and a fixed slope (dashed lines) are fitted to the data. The corresponding thresholds are estimated from the intersection point with respect to the threshold obtained from the full fit. The vertical lines indicate the excitation power densities for later time-resolved measurements. In the lower panel, the relative error for the performed fit is plotted, for clarity every third point is missed out.

Besides, the threshold is estimated from the data by using linear fits to the points at lower and higher excitation power densities. Two approaches are used for the linear fit. The first strategy is by fitting linear functions and allowing the slope to vary. In that case, the slope of the non-linear regime is ~1.8 and nearly linear at high excitation power densities ~1.0. The estimated threshold is 1.4 times higher than the threshold determined from the full fit (15.8 mW cm^{-2}). The second strategy is by fixing the slopes of the linear functions to 2 and 1, in this case the threshold is 0.9 times the threshold from the full fit (10.2 mW cm^{-2}). In both cases, the resulting thresholds are deviating from the threshold obtained from the fit of the entire intensity dependence.

3.2.1 Time-dependent analysis

For the verification by time-resolved data the laser diode is time modulated and the rising, as well as falling edge, are measured at seven excitation power densities indicated by vertical lines in the top panel of Figure 3.7. These are in multiples of the previously obtained steady-state threshold: 0.1, 0.6, 1.0, 2.5, 22.0, 95.3, 226.7. Figure 3.8A shows the rising edge of the upconversion emission at $473\,\mathrm{nm}$. To guarantee that steady-state is reached a pulse duration of 10 ms is chosen.

A global non-linear least square fit by using the rate equations defined previously is performed for the rising and the falling edge of every excitation power density. From the static quenching of the phosphorescence signal in Figure 3.6 the transfer efficiency close to unity is assumed $\phi_{\mathrm{DET}} = 1$. The sensitizer triplet lifetime is obtained from an additional time-resolved measurement of the phosphorescence signal in a solution that contains only Pd-TBP, $\mathrm{k_D^S} = 3570\,\mathrm{s^{-1}}$. The remaining parameters that are varied in the global fit are the emitter triplet decay rate $\mathrm{k_E^D}$ and the collision frequency γ. Figure 3.8 shows the measured data of the rising edge of the upconversion emission. The rising edge behaves very similarly to the derive analytic equations that are plotted in Figure 3.5.

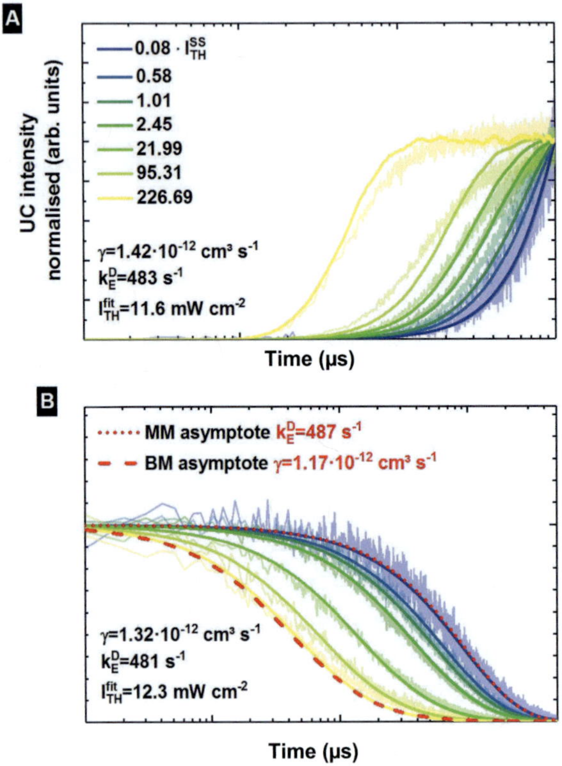

Figure 3.8: A) The upconverted emission intensity at 473 nm (translucent, solid lines) as a function of time for excitation at various excitation power densities (given as multiples of the 11.3 mW cm^{-2} threshold determined previously by the steady-state analysis. A global fit to the rate equations is shown by the solid lines (solid lines). Best fit values extracted from the global analysis for the emitter triplet decay rate k_E^D and the bimolecular recombination constant γ are shown. B) Analogous analysis for the falling edge transients of the upconversion emission intensity

The best fit is performed with the inset parameters, $\mathrm{k_E^D} = 483\,\mathrm{s}^{-1}$ and $\gamma = 1.42 \cdot 10^{-11}\,\mathrm{cm}^3\,\mathrm{s}^{-1}$. From equation (3.17), an upconversion threshold is calculated, $\mathrm{I_{TH}^{fit}} = 11.6\,\mathrm{mW\,cm}^{-2}$. Figure 3.8B shows the data and the global fit for the falling edge. The initially excited state concentrations for the falling edge are found by calculating the steady-state values from the rate equations. The extracted values for the emitter triplet decay rate, $\mathrm{k_E^D} = 481\,\mathrm{s^{-1}}$ and the collision frequency, $\gamma = 1.32 \cdot 10^{-12}\,\mathrm{cm^3\,s^{-1}}$ are close to the values obtained from the rising edge. The corresponding threshold is $\mathrm{I_{TH}^{fit}} = 12.3\,\mathrm{mW\,cm^{-2}}$. For completeness, the asymptotic monomolecular falling edge is fitted to the falling edge of the smallest excitation power density. The so obtained emitter triplet decay rate is $\mathrm{k_E^D} = 481\,\mathrm{s^{-1}}$ and corresponds to the value obtained from the global fit.

3.2.2 Summary of experimental verification

The demonstrated verification shows that the predictions based on the used rate equations match excellent to the experimental data for the perylene/Pd-TBP system. The small deviation in the collision frequency γ between the steady-state and the time-resolved data could be due to thermal effects. The reason is closely related to the fact that the collision frequency is proportional to the diffusion constant:

$$\gamma = 8\pi D R_0, \tag{3.37}$$

where D is the diffusion constant and R_0 is the interaction radius of the collision partners. In the case of a continuous excitation a thermal gradient is induced that increase the diffusion constant in the solution. In the case of the modulated excitation, the heating is lower and so is the diffusion constant. Hence, the threshold obtained from the steady-state experiments is smaller than the threshold obtained from the time-resolved measurements.

Exciting features can also be found by looking onto the dynamic quenching of the phosphorescence in the presence of perylene. In the analysis of the time dependent data a ϕ_{tr} close to unity is estimated from the static quenching of the phosphoresence. However, a closer look at the time dynamics of the phosphorescence in the presence of perylene unveils a process that is not covered in the discussion. Figure 3.9 shows the transient of the phosphorescence for a pure Pd-TBP solution. The transient shows a monoexponential decay with a lifetime of $\tau_S^D = 280\ \mu s$. The corresponding sensitizer triplet decay rate is $k_S^D = 3570\ s^{-1}$. Additionally, the transient of the phosphorescence of the TTA-UC solution is recorded. Instead of a sharp edge, that would correspond to $\phi_{tr} = 1$, a fast and a long-lived component could be observed. The long-lived component has a lifetime of $\tau_E^D = 2083\ \mu s$ and a corresponding decay rate of $k_E^D = 480\ s^{-1}$. That coincides with the emitter triplet decay rate that is found previously, suggesting that there is a small portion of emitter triplets that transfer back onto Pd-TBP.

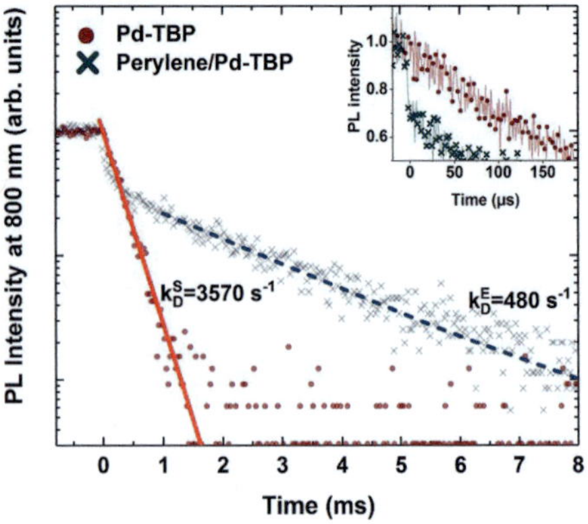

Figure 3.9: The transients of the phosphorescence signal at 800 nm are shown after a continuous wave excitation. The solution containing only Pd-TBP (circles) is fit to a mono-exponential function (solid line), and a decay rate is obtained. The phosphorescence in the Perylene/Pd-TBP solution (crosses) shows a prompt decay right after the laser is turned off, also shown in the inset plot in the top right corner. The long living decay is fitted to a mono-exponential function (dashed line) and a corresponding decay rate is obtained.

Monguzzi *et al.* had discussed the back-transfer on the UC-PLQY. Their data shows that at high excitation power densities the intensity dependence of the UC intensity changes below one because of the back-transfer. [63]

However, this is not observed mainly because the sensitizer concentration ($c = 5 \cdot 10^{-5} M/L$) used is low enough.

Despite the demonstrated back-transfer, theory and experiment agree well. Based on these findings, a new hypothesis is now stated for a non-uniform mixed solution but for a bilayer that is made from one sensitizer and one emitter layer.

3.3 Rate equation formulation for anisotropic one-dimensional systems

The previous part dealt with a system in which the sensitizer and emitter molecules are distributed uniformly. The next step is to extend the theory onto the case in which the sensitizer and emitter are two separate layers that are connected by one interface. The coupling between emitter and sensitizer happens at this interface. The excited sensitizer triplets must diffuse to the connecting interface at which they transfer to the emitter layer and in which TTA-UC can happen. Although the energy will move from one molecule to another it is assumed that the energy diffusion length is large compared to the distance between two sensitizer or emitter molecules. Therefore, energy transport can be modeled by Fick's diffusion equation. Further, it is assumed that the layer thicknesses are small compared to the other two dimensions so that a one-dimensional formulation is valid. Assuming $\phi_{\text{ISC}} = 1$ the triplets in the sensitizer layer are governed by the following diffusion equation:

$$\frac{\partial S_T}{\partial t} = D_S \frac{\partial^2 S_T}{\partial x^2} + G(x) - k_S^D S_T - k^{\text{DET}}(x) S_T, \qquad (3.38)$$

here D_S is the diffusion constant of the sensitizer triplets, $G(x)$ is the generation rate according to Lambert-Beer's law $G(x) = G_0 \exp(-\alpha x)$ and $k^{\text{DET}}(x)$ is the transfer rate which is restricted to the coupling interface. To reduce the complexity of the problem the generation rate is reduced to a constant that corresponds to a small absorption across the sensitizer layer. Additionally, the transfer process is included as boundary condition at the interface $x = d_S$:

$$G(x) = G_0, \tag{3.39}$$

$$k^{\mathrm{DET}}(x) \rightarrow S_T(x = d_S) = 0, \tag{3.40}$$

here $G_0 = \alpha I_0$ and d_S is the thickness of the sensitizer layer. Further, we assume that steady-state is reached and reflective boundary conditions are fulfilled at the remaining interface $\left.\frac{\partial S_T}{\partial x}\right|_{x=0} = 0$. With these assumptions, the partial differential equation (3.38) reduces to the following differential equation:

$$D_S \frac{\partial^2 S_T}{\partial x^2} = -G_0 + k_S^D S_T \tag{3.41}$$

Together with the described boundary condition, the solution is:

$$S_T(x) = \frac{G_0}{k_S^D}\left(1 - \operatorname{sech}\left(\frac{d_S}{L_S}\right)\cosh\left(\frac{x}{L_S}\right)\right),$$

where $L_S = \sqrt{D_S/k_S^D}$ is the diffusion length of the sensitizer triplets. The number of sensitizer triplets that moved to the emitter layer is the difference between of the number of sensitizer triplets generated and the number of sensitizer triplets that remain:

$$N_S^{\mathrm{DET}} = A G_0 L_S \tanh\left(\frac{d_S}{L_S}\right), \tag{3.42}$$

where A is the beam area. For thicknesses much smaller than the diffusion length $d_S \ll L_S$ this simply reduces to:

$$N_S^{\mathrm{DET}} = A G_0 d_S. \tag{3.43}$$

The result is intuitively explainable, as long as the thickness of the sensitizer layer is thinner than the diffusion length L_S, most of the generated sensitizer triplets can move towards the interface and are transferred to an emitter layer. For the modeling of the emitter layer, it is assumed that the diffusion length of the emitter triplets is much longer than the emitter layer thickness $L_E \gg d_E$. Therefore, the transferred triplets are spread uniformly across the emitter layer $N_S^{\mathrm{DET}}/(Ad_E)$. Hence, the equation that describes the emitter triplets at steady-state is:

$$0 = D_E \frac{\partial^2 E_T}{\partial x^2} + \frac{N_S^{\mathrm{DET}}}{Ad_E} - k_E^D E_T - \gamma E_T^2, \tag{3.44}$$

where D_E is the diffusion constant for the emitter triplets and d_E is the emitter layer thickness. It is assumed that there is no back-transfer of emitter triplet to the sensitizer layer. That means that reflecting boundary conditions have to be used at $x = 0$ and $x = d_E$. This means that the steady-state solution is a constant function and so the diffusion term is zero. The intensity dependence follows the derivation in section 3.1.1:

$$\tilde{I}_{\mathrm{UC}} = \left(\sqrt{1 + 4\frac{G_0}{G_{TH}(d_S, d_E)}} - 1\right)^2 . \tag{3.45}$$

The threshold, in this case, is a function of the sensitizer and emitter layer thickness:

$$I_{\mathrm{TH}} = \frac{1}{\alpha}\frac{1}{\phi_{DET}}\frac{k_E^{D^2}}{\gamma}\frac{d_E}{d_S}.$$

3.4 Summary

In this chapter the equations for TTA-UC were solved for the steady-state and the time-dependent case. The found solutions were verified by using a model system. Based on the agreement between theory and experiment a hypothesis for the threshold was made for a bilayer system. The so derived threshold is now not only depending on static molecular parameters but can be tuned by varying the thickness of the sensitizer and emitter layer thicknesses. The additional dependence on the sensitizer layer thickness d_S can be interpreted as an increase of the overall absorbed photons by the system. The dependence on the emitter layer thickness d_E suggests that thinner emitter layers will increase the collision frequency γ. Both, a thick sensitizer layer and a thin emitter layer increase that TTA efficiency at a given excitation power density and therefore lower the threshold. Yet, it is important to state that this behavior can only be observed if and only if the exciton diffusion length is larger than the sensitizer and emitter layer thickness. In the situation, where the sensitizer layer thickness is larger than the exciton diffusion length the threshold will rise as the sensitizer layer gets thicker. This implies, that for a successful proof of this theory a material has to be found that allows very good exciton diffusion length. In our case, we assumed that these can be SURMOFs, as they provide a very ordered and structured alignment of molecules reducing the transport losses coming from disorder.

3.4.1 Contributions to this chapter

Listed are the contributions from persons I have worked with in form of practical help or constructive ideas.

Oldenburg, M. (Author):

- Modelling and derived solutions to the models.

- All spectroscopic measurements.

- Figures and photographs.

Kiseleva, N. and Turshatov, A.:

- Preparation of the samples (upconversion solutions).

Busko, D.:

- Experimental setup for steady-state emission measurements and time-resolved measurements.

Oldenburg, M., Howard, I. A. and Kim, D.:

- Implementation of the global fit code into Matlab and that is further mentioned in the master thesis of Kim, D. [89]

Howard, I. A., Richards, B. S. and Oldenburg, M.

Elaborating the bilayer idea for solid-state upconversion layers based on SURMOFs.

4 Optimization of an emitter layer

This chapter is based on parts of the following publication:

M. Oldenburg, A. Turshatov, D. Busko, M. Jakoby, R. Haldar, K. Chen, G. Emandi, M. O. Senge, C. Wöll, J. M. Hodgkiss, B. S. Richards and I. A. Howard, *Enhancing the photoluminescence of surface anchored metal–organic frameworks: mixed linkers and efficient acceptors,* Phys. Chem. Chem. Phys., 2018, DOI: 10.1039/c7cp08452h.

The goal of the entire thesis is to make a bilayer structure from SURMOFs which under illumination performs TTA-UC. One requirement for efficient upconversion is to enhance the photoluminescence quantum yield (PLQY) of the emitter layer and at the same time keep the energy transport properties as good as possible. In this chapter, the energy transport properties in the emitter SURMOF based on diphenylanthracene are analyzed. Based on this the low PLQY of 1% is explained. Two strategies are analyzed to enhance the PLQY. In the first approach, SURMOFs are fabricated from a mix of an emissive linker with an optically-inert linker of equal length, diluting the emissive linker while maintaining the SURMOF structure. In the second approach, the behavior is explored wherein energy-accepting guest chromophores are infiltrated into SURMOF that acts as energy valves to release the energy in the form of luminescence before it relaxes.

4.1 Structure analysis of ADB SURMOF

The decision on DPA based SURMOF is motivated by the fact, that together with Pt-OEP it is the benchmark leading system for organic upconversion. [81] For the formation of the emitter SURMOF structure, two carboxylic acids are attached to the phenyl rings of DPA, the resulting molecule is the ditopic carboxylate 4,4'-(anthracene-9,10-diyl)dibenzoate (ADB). ADB is coordinating two zinc atoms to a paddle-wheel structure and results in a SURMOF with P_4 symmetry in the structural class known as SURMOF-2.[39] The structure is illustrated in Figure 4.1. From XRD measurements a distance into (001) and (100) direction of 2.2 nm, and 0.6 nm in (010) direction could be estimated by using the Bragg condition:

$$2d\sin(\theta) = n\lambda, \tag{4.1}$$

where d is the distance between two lattice surfaces, θ the angle of the observed peak, n the order of corresponding peak and λ the wavelength of the x-rays. The out-of-plane XRD in Figure 4.1C shows five orders of the (001) peak. The absence of other orientations shows that the ADB SURMOF crystals have a single orientation with respect to the substrate surface. The in-plane XRD in Figure 4.1C shows peaks for the (100) orientation together with a single peak of the (010) orientation. A wrong interpretation would be to assume that this means that the SURMOF forms a single crystal thin film. Hence, crystalline domains that are rotated with respect to the substrate's normal are not contributing to the XRD. Figure 4.2A shows the SEM image of the surface of a spray casted ADB SURMOF. The image shows flakes that are upstanding on top of the substrate, however that are rotated randomly against each other. In contrast to this, Figure 4.2B shows the SEM image of a spin-coated ADB SURMOF that has a much smoother surface. The difference in morphology for spray coated and spin-coated ADB SURMOFs could be explained by the slower growth rate in the case of spin coating allowing the formation of a denser film. By using the Debye-Scherrer equation: [91]

$$\Delta(2\theta_0) = \frac{K\lambda}{L \cdot \cos(\theta_0)}, \tag{4.2}$$

where $\Delta(2\theta)$ is the full width half maximum of the first order peak, $K \sim 1$ is the Scherrer form factor, L the crystallite size and θ_0 the position of the first order peak the growth rate can be estimated. The growth rate for spray coating is ~15 nm per cycles; in the case of spin coating the growth rate is ~2.2 nm. The later would correspond to single molecular layer deposited after each cycle. [92]

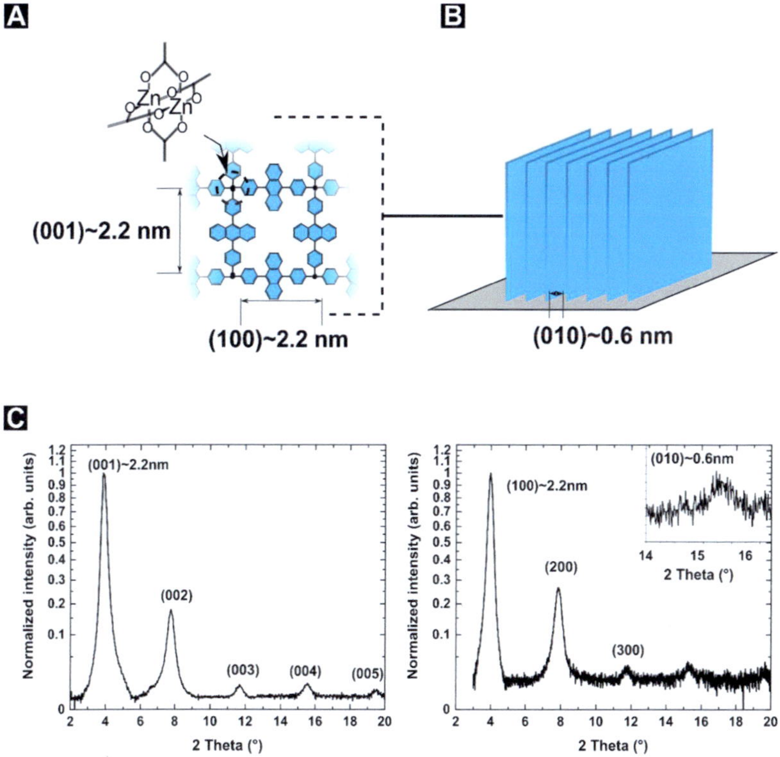

Figure 4.1: A) The structure of a single ADB sheet is illustrated schematically. The ADB linker is coordinating two zinc atoms to form a paddle-wheel structure that is shown in the upper left corner. The distances into (001) and (100) direction are shown by arrows together with the corresponding lattice distance. B) A schematic of several ABD SURMOF sheets is shown that stand on a substrate (grey area) with an intralayer distance into (010) direction. C) Left: XRD is shown for the out-of-plane measurement configuration of an ADB SURMOF on silicon. Five orders of (001) peak show very ordered alignment of the ADB SURMF. Right: In-plane XRD for an ADB SURMOF on silicon showing three orders of (100) peak. Inset: Magnification of the region in which the first order of the (010) peak can be seen. This figure is adapted from Oldenburg et al. [92]

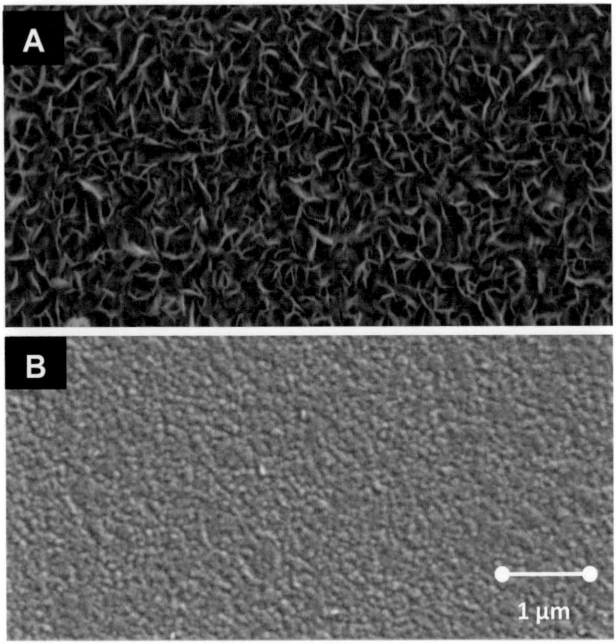

Figure 4.2: SEM images (with a magnification of 10000) of ADB SURMOFs synthesized on soda lime glass. A) ADB SURMOF produced by spray coating. B) ADB SURMOF produced by spin coating. The scale bar of 1µm applies for A) and B). This figure is adapted from Oldenburg et al. [92]

4.2 Photoluminescence properties

The close distance into (010) direction leads to interactions between the ADB chromophores. The resulting aggregate-like behavior limits the PLQY of the ADB-based SURMOFs. A PLQY is measured of less than 5% for a SURMOF made solely of ADB, whereas the PLQY of the linker molecule in ethanol solution is nearly 100%. The main loss is attributed to the formation of non-radiative aggregate states. The separation of the ADB molecules from each other, as in a solution, would lead to a reduction of these losses. In order to proof this hypothesis and to understand the transition from single ADB molecules towards densely packed ADB SURMOFs the structure shown in Figure 4.1A is modified. The ditopic carboxylate p-terphenyl-4,4'-dicarboxylate (TPC) that has nearly the same length as ADB is used as an inactive linker to separate the ADB molecules from each other. TPC has its absorption peak at 275 nm and emits around 350 nm (see Figure 4.3A). For this reason, TPC is referred to as an optical inactive linker. Figure 4.3B shows the schematic structure of a SURMOF made from TPC and ADB. Three SURMOFs from a solution containing a mixture of ADB and TPC linkers are fabricated. As ADB and TPC are ditopic carboxylates Zn2+ "paddle-wheel" metal centers are coordinated by these linkers forming two-dimensional sheets standing perpendicular to the substrate (as already discussed for ADB).[93-95] The closest proximity between two linkers in the mixed SURMOF structure is in the direction between sheets (010) and is 0.6 nm, as it is the case for the ADB SURMOF. This strategy of the application of two similar or equally-sized linkers within a MOF is shown by Dhakshinamoorthy et al. [96]

Four SURMOFs with different molar ratio between ADB and TPC are prepared: i) one is made purely from ADB; ii) one with an ADB to TPC ratio of 1:1; iii) one with an ADB to TPC ratio of 10:1; and iv) one consisting of pure TPC. For the analysis of the time dependent evolution of excitons in each SURMOF time-resolved PL is used as it is shown in Figure 4.4A. By using a streak camera, the emission spectrum from 350 to 600 nm is analyzed within a time range from 0 to 10 ns. All three samples containing ADB are excited at 350 nm with a femtosecond pulsed laser. [92]

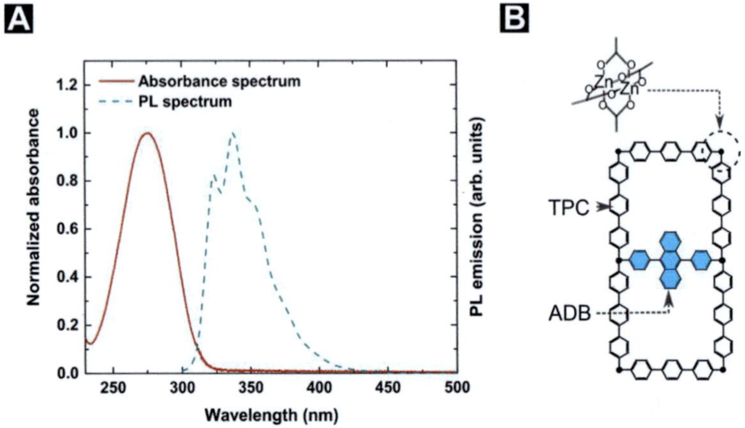

Figure 4.3: A) The absorbance (solid line) and PL spectrum (dashed line) of TPC. B) A SURMOF made from TPC and ADB (showed by arrows). This figure is partially based on a figure in Oldenburg's et al. publication. [92]

In Figure 4.4B the spectrum of the emission is shown. As can be seen it changes with time that shows that there are two excited-states that have two different lifetimes. For guiding the eye a green dotted line is integrated that is a spline through the red points. The red dots correspond to a Gaussian fit at certain time intervals. For all samples, the emission shifts to the red with time: the pure ADB SURMOF has a shift of 40 nm (0.22 eV), while the shifts for the SURMOFs having 50% and 10% ADB are 45 nm (0.24 eV) and 50 nm (0.26 eV), respectively. This shows that the nature of the two states responsible for the emission is nearly unchanged in the three samples. [92]

For understanding the dynamics of the excited-states, the time-resolved data are factored by applying the method of multivariate curve resolution together with a constrained alternating least squares algorithm (MCR-ALS) with non-negativity constraints on both the spectra and kinetics. [97] By doing so, the method provides a PL spectrum and population decay profile for both excited state; the spectra and kinetics are shown in Figure 4.4B and C, respectively. The MCR-ALS analysis shows that there is a short-lived component whose spectrum and kinetics are plotted on the left side of panel Figure 4.4B and C (in blue colors) and a long-lived state whose spectrum and kinetics are plotted on the right side of the panels (in red colors). The PL emission for the short-lived component has an emission maximum around 445 nm that fits good to the emission spectrum of the ADB solution that is plotted in Figure 4.4B as a filled curve (with a maximum around 440 nm). Therefore, the blue emission corresponds to the monomer-like state in the SURMOF. The second state with a maximum at 510 nm, is the same for all three samples. This state arises from the interaction of many ADB chromophores in the SURMOF. [92]

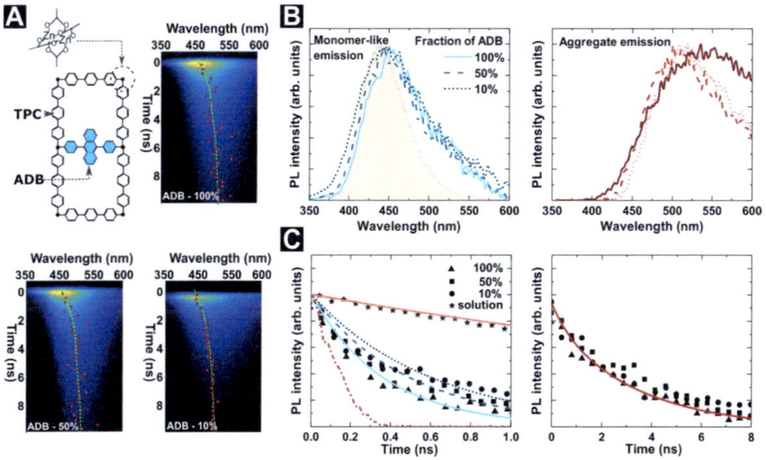

Figure 4.4: A) Drawing of one sheet of the mixed SURMOF fabricated from a zinc center coordinated by 4,4'-(anthracene-9,10-diyl)dibenzoate (ADB) and p-terphenyl-4,4'-dicarboxylate (TPC); and streak camera measurements of SURMOFs made from 100%, 50%, and 10% of ADB linkers excited at 350 nm. The red dots show the emission peak for certain time intervals and the green dashed curve is a guide to the eye provided by a spline function through the red dots. The graphs in B) and C) show the PL emission spectrum and decays, respectively, of the two states responsible for the emission. B) The emission spectrum of the monomer-like (left) and aggregated state (right) for the three SURMOFs. The spectrum of the ADB linker in ethanol solution is shown as a filled curve. C) The kinetics of the monomer-like and aggregate state of each sample. This figure is copied with permission from Oldenburg et al. [92]

In Figure 4.4C the lifetime profiles for the monomer and aggregate components are plotted. From a single-exponential fit the lifetimes for both states were found. The lifetime of the monomer-like state rises as the concentration of TPC is increased $\tau_{100\%} = 0.37 \pm 0.01$ ns to $\tau_{50\%} = 0.50 \pm 0.01$ ns to $\tau_{10\%} = 0.61 \pm 0.02$ ns. In contrast, lifetime of ADB in solution is 4.3 ± 0.01 ns. From this analysis it is concluded that the lifetime of the bright state is reduced as the amount of ADB is increased. In comparison to the short lived-state the long lived state remained the same for all samples, 3.3 ± 0.3 ns. [92]

The monomer lifetimes correlate with the PLQY measured for the three different SURMOFs. The plot in Figure 4.5A shows, that maximum of the PL emission is shifting to the blue and the brightness increases as the amount of ADB is reduced (maximum of the PL spectrum is at 450 nm for the 100% ADB sample, 440 nm for the 50%, and 437 nm for the 10% sample). The measured average internal PLQYs are 15±2, 8±4 and 4±3% for the samples made with 90, 50, and 0% of TPC, respectively. The averages and error estimations are based on three samples that were made independently for each ratio. Further, the internal PLQY is defined by the ratio of the number of photons emitted to the number of photons absorbed. From the close proximity of the peak of the steady-state emission (shown in Figure 4.5A) and the peak of the monomer-like emission found from the time-resolved analysis in Figure 4.4B, it can be stated that the PLQY of the short living excited state is much higher than the PLQY of the long living and red shifted state. Hence, the total PLQY is coming from the emission of the monomer-like state. Therefore, the PLQY can be calculated as follows:

$$\eta_{SURMOF}^{int} = \eta_{solution}^{int} \cdot \frac{\tau_{monomer}}{\tau_{solution}}, \tag{4.3}$$

here $\eta_{SURMOF}^{int}, \tau_{monomer}$ and $\eta_{solution}^{int}, \tau_{solution}$ stand for internal PLQY and PL lifetime of the monomer-like state in the SURMOF and solution. Using the values for the lifetimes presented above and the assumption that

$\eta_{solution}^{int} \approx 100\%$, an estimation of the internal PLQY can be performed for all three samples: 14±0.5, 12±0.3, and 8±0.2% for the 10, 50, and 100% ADB samples, respectively. The agreement between the directly measured PLQY and the from the lifetime decrease of the monomer-like state concluded PLQY indicates that the formation of aggregate states in the SURMOF is the main PL loss mechanism within the SURMOF. [92]

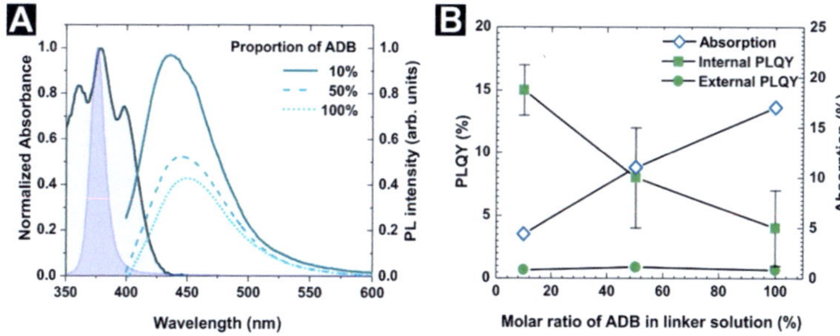

Figure 4.5: A) PL emission spectra of the SURMOFs with different ADB fractions, the absorption spectrum of the ADB linker and the spectrum of the LED excitation. The PL spectra are weighted with the largest internal PLQY. B) The internal PLQY, sample absorption, and external PLQY as a function of ADB linker molar ratio in the linker solution. This figure is copied with permission from Oldenburg et al. [92]

Figure 4.5B shows the effect of the changing molar ratio of the ADB linker on the *external* PLQY, that is defined as the number of emitted photons divided by the number of photons *incident* on the sample. Hence, the external PLQY is equal to the internal PLQY multiplied by the fraction of photons absorbed. Figure 4.5B shows that the absorption decreases as the fraction of ADB within the SURMOF is reduced. This indicates that the external PLQYs are similar and below 1% for all three ratios, namely 0.66±0.1, 0.88±0.44, and 0.68±0.51% for the 10, 50, and 100% ADB samples, respectively. [92]

In the next part, another strategy is proposed to increase the internal PLQY and at the same time keep the absorption of the sample at nearly the same level.

4.3 Coumarin 343 as an efficient energy acceptor

In this strategy, the ADB SURMOF is infiltrated by a varying amount of one of two coumarin derivatives that are mixed into the linker solution. Two dyes were used, coumarin 343 (C343) and coumarin 334 (C334). From the overlap between the PL emission spectrum of the ADB SURMOF and the absorption spectrum of C343 and C334 it is concluded that energy transfer between SURMOF and infiltrated dye will take place. The PL spectra of the monomer and aggregate emission together with the absorbance spectra of C343 and C343 are shown in Figure 4.6. [92]

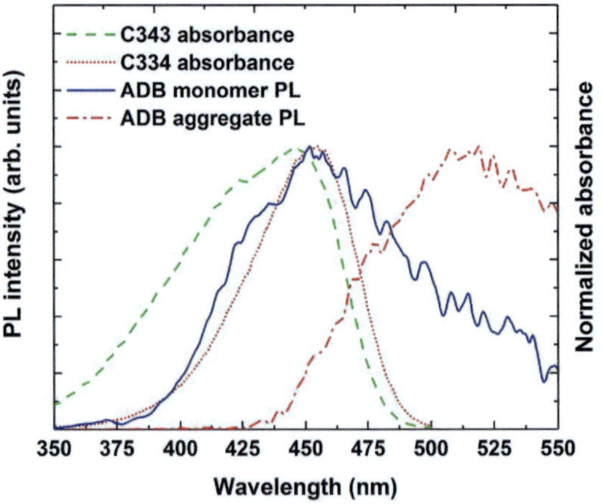

Figure 4.6: The emission spectrum of ADB the monomer and aggregate state are shown together with the absorbance spectra of coumarin 343 (C343) and coumarin 334 (C334) in ethanol. This figure is copied with permission from Oldenburg et al. [92]

The main difference between C343 and C334 is that they have two different functional groups at the molecular position 10 (C343 has a carboxylic group while the C334 has a methyl ketone group). It is assumed that the carboxylic group leads to strong interaction within the SURMOF. [17] This is partially shown by infrared reflection-absorption spectroscopy (IRRAS). Figure 4.7A shows the infrared reflection-absorbance spectra of a pure and an ADB SURMOF loaded with C343 grown on MHDA/Au substrate. The first sign that C343 is loaded in the ADB SURMOF is the broad band around 3400 cm^{-1}. This is the stretching of OH groups coming from C343. This stretching is absent in the pristine ADB SURMOF. Figure 4.7B shows the bands at 1590 cm^{-1} and 1420 cm^{-1} that are assigned to the asymmetric and symmetric stretching of COO^-, showing the formation of a zinc paddle-wheel structure with bidentate ligands as the difference of both bands it between 140 and 200 cm^{-1}. This shows that the SURMOF is intact in the case of pure ADB SURMOF and in the case of loaded ADB SURMOF. Figure 4.7C Shows the presence of a C-N stretching band around 1130 cm^{-1} that comes from C343 molecule, this band is absent in the pristine ADB spectrum. [98] Additionally, Figure 4.7D shows the absence of free carboxylic group vibrations in both, pristine ADB SURMOF and loaded ADB SURMOF that might indicate that C343 is not only loaded into the ADB SURMOF but actually is part of the whole SURMOF structure. Figure 4.7E shows the absence of free carboxylic groups that can be attributed to strong hydrogen bonds formation, which form sp^3 C-H bands around 2860, 2845 and 2835 cm^{-1} in the case of the loaded ADB SURMOF or that C343 is reacting during the synthesis of the SURMOF. [92]

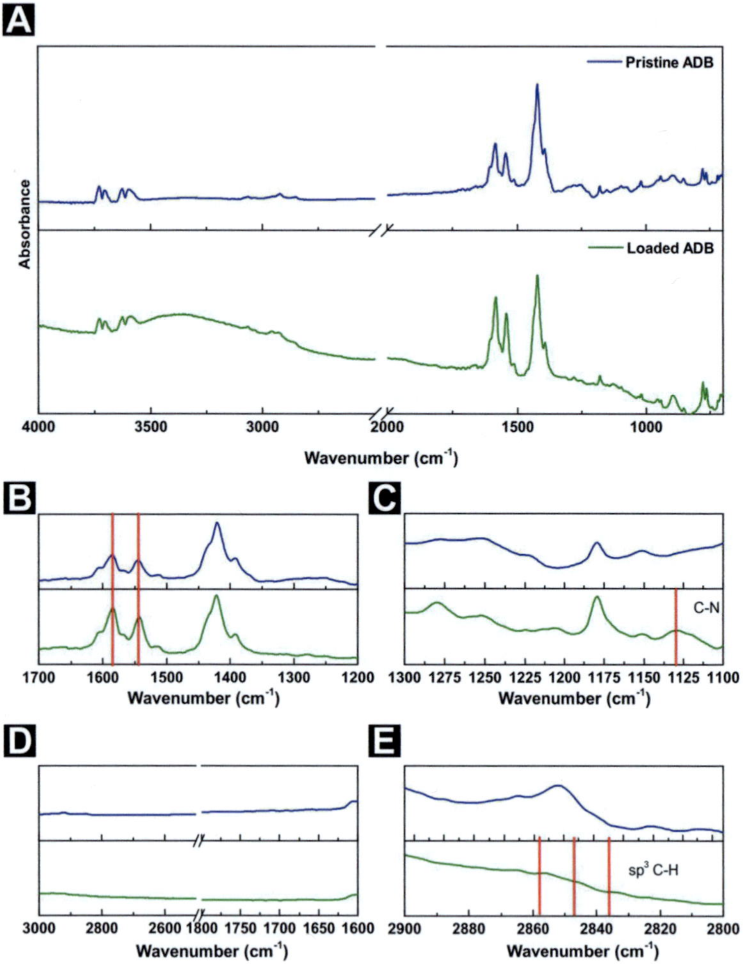

Figure 4.7: Infrared reflection-absorption spectrum (IRRAS) of an ADB SURMOF grown on MHDA/Au substrate (upper panel, blue line) and ADB SURMOF loaded with coumarin 343 (lower panel, green line). The vertical solid lines are marking corresponding bands and are described in the main text. This figure is copied with permission from Oldenburg et al. [92]

The IRRAS analysis shows that C343 is a part of the whole SURMOF structure. Yet, evidence for a non-covalent bonding of the C343 molecule within the SURMOF is shown by washing a C343-containing SURMOF via immersion in a solvent (such as toluene and ethanol) for three days. The use of two different solvents is due to different interactions between the solvent and the C343. In the case of toluene, the benzene ring should lead to stronger interactions compared to ethanol. The difference is shown in the inset of Figure 4.8. Although some of the C343 is washed out (leading to a slight blue-shift of the PL emission spectrum shown in Figure 4.8) most stays within the SURMOF. This leads to the conclusion that C343 molecules within the SURMOF are either weakly bound by hydrogen bonds or strongly bound by covalent bonds. [92]

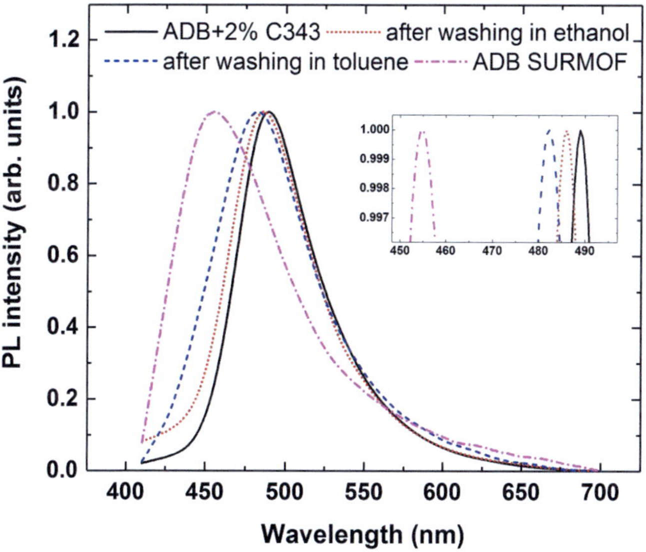

Figure 4.8: Two ADB SURMOFs with 2% of coumarin 343 in the linker solution are synthesized and the emission spectrum is recorded (black). Both samples are immersed for 3 days in ethanol (red curve) and toluene solution (blue curve). As a comparison the spectrum of pristine ADB SURMOF (purple) is shown. The inset shows a magnification of the peak position of every curve. This figure is copied with permission from Oldenburg et al. [92]

Figure 4.9A shows a drawing on how C343 and C334 could be lying within the ADB SURMOF. C343 is drawn as a green molecule that interacts with the SURMOF by the carboxylic group (the case for strong covalent bonding is not shown). C334 is drawn as a red molecule in Figure 3a, which does not form any bonds in the SURMOF. [92]

Samples with nine different molar ratios between the coumarin dyes and the ADB linker are made (30, 20, 10, 5, 2, 1, 0.1, 0.02, 0.005% molar ratio of the guest molecule to the linker). Figure 4.9B illustrates the absorption and emission spectra of each of these SURMOFs. Additionally, the absorption and emission spectra of C343 and C334 solutions as filled curves are shown. The analysis of the samples that were infiltrated by the C343 molecule show a clear correlation between the amount of C343 and the PL spectrum. The emission of with the lowest amount of C343 (0.005%) is mainly coming from the ADB emission. This leads to an PL spectrum that is blue-shifted with respect to that of the C343 in solution. Yet, by increasing the C343 concentration the absorption peak at 440 nm grows and the PL spectrum shifts towards the red. The red shift of the PL emission spectrum with increasing concentration can be due to two reasons: i) the increasing reabsorption of the blue C343 emission due to the growth of the absorption at 440 nm and ii) due to a solvatochromic effect induced by either the hydrogen-bonding of the C343 chromophores or covalent bonding (as has been observed for solvent environments).[92, 99]

In comparison, the ADB SURMOFs with C334 (shown in red in Figure 4.9B) show a significantly different behavior. Firstly, there is no appearance of an absorption peak at 440 nm in the SURMOF at high molar ratios of C334. This is due to less uptake of C334 in the ADB SURMOF. It is assumed that a lot of C334 is washed out during the synthesis and therefore not much influence on the PL spectrum can be seen. This is also showed in Figure 4.9C. Here, the wavelength position of the PL maximum for SURMOFs with C343 and C334 as a function of the molar ratio is shown. In the case of C343, clear evidence of a controlled infiltration into the SURMOF is shown. [92]

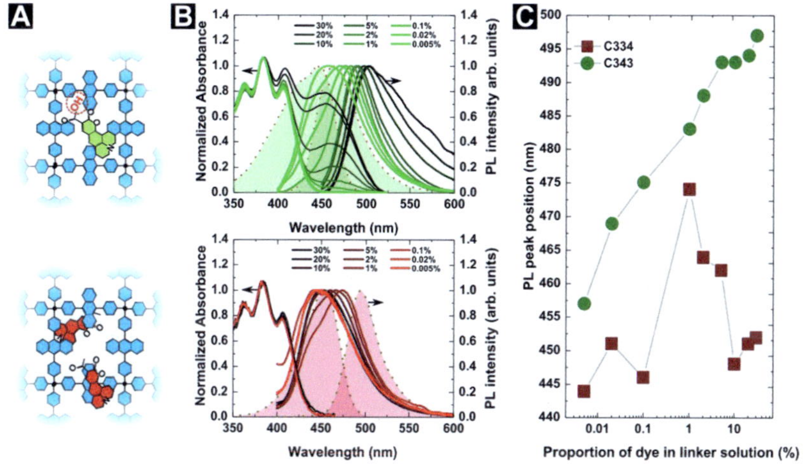

Figure 4.9: A) Sketch of ADB SURMOFs with coumarin 343 (C343, green colored molecule) (upper drawing) and with coumarin 334 (bottom drawing, red colored molecule). The dashed red line shows the hydrogen bond interaction of C343 with the ADB SURMOF. B) Plotted are the absorption and emission curves of C343 and C334 solutions, both as filled curves. The solid curves show the absorption and emission spectrum of ADB SURMOFs with different concentration of C343 (top) or C334 (bottom). C) The peak maximum of the PL emission spectra is plotted as a function of the molar ratio of C343 and C334 in the linker solution. This figure is copied with permission from Oldenburg et al. [92]

The data sets of ADB SURMOFs with C343 and C334 show that a controlled amount of C343 into the ADB SURMOF can be introduced. A comparison study shows how a dropcasted C343 solution behaves in contrast. For this a C343 solution is dropcasted on a neat ADB SURMOF and after the C343 solution dried, the SURMOF is washed with ethanol. The PL spectrum is shown in Figure 4.10A. Interestingly, the red curve of the dropcasted sample correspond very well with the averaging over all PL emission spectra that are recorded from the C343 loading described previously (as shown in Figure 4.10B). This might be interpreted as a non-uniform distribution of C343 molecules in the dropcasted sample. This interpretation agrees well with the measured PLQY that is only 7%. [92]

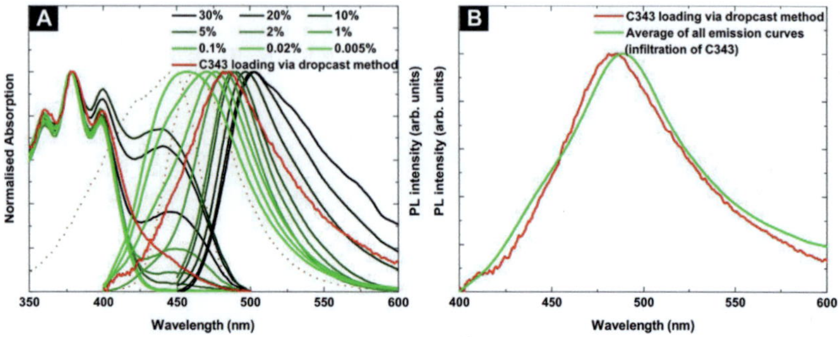

Figure 4.10: A) Normalized absorbance and PL emission spectra of different ADB SURMOFs infiltrated by C343 with different molar ratios (9 shades of green). Compared to this the absorbance and PL emission spectrum of an ADB SURMOF dropcasted and washed with a C343 solution (red). Panel B) shows the emission spectrum of the drop-casted sample (red) and the average emission of all 9 samples that are loaded with C343 during the fabrication. This figure is copied with permission from Oldenburg et al. [92]

Based on the earlier studies a correlation of the PLQY and the molar ratio of C343 and C334 is plausible. The PLQY of C343 in solution is 63%.[100] Figure 4.11A shows the dependence of the PLQY as a function of the molar ratio of C343 in the linker solution. ADB SURMOFs with different concentrations of C343 (C343@ADB) prove a remarkably interesting behavior: when the molar ratio of dye is reduced from 30% to 1% the PLQY increases and reaches a maximum of 50±2%. As the amount of C343 is reduced below 1%, the PLQY starts to decrease, reaching 28% for a molar ratio of 0.005%. This behavior can be easily understood by a finite-range the energy transfer within the ADB SURMOF and the aggregation effect between the C343 molecules formed at high concentrations. As the molar ratio decreases the average distance between the C343 dyes increases. Therefore, the PLQY is reduced as the excited states from the ADB SURMOF cannot react C343 dyes anymore. In contrast, at very high C343 concentrations the excitons reach a C343 dye, however, due to close proximity of the C343 dyes the reabsorption is increased and aggregate states are formed leading to a reduction of the PLQY. Figure 4.11B shows a photography of 5 SURMOF with 5 different molar ratios of C343 in the linker solutions showing that the SURMOF where the sample with a molar ratio of 1% C343 is the brightest. [92]

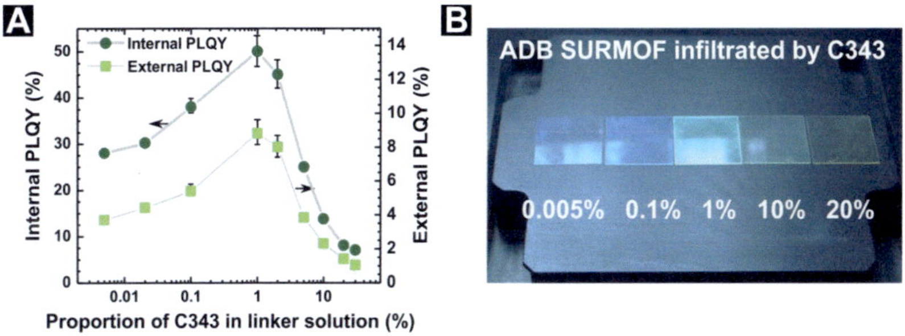

Figure 4.11: A) The internal and external PLQY of ADB SURMOFs infiltrated by C343 as a function of the concentration of the dye in the ADB linker solution. Samples with 2%, 1% and 0.1% of C343 are prepared and investigated three times to estimate the error, which is shown by the error bar. B) Photo of five ADB SURMOFs samples with C343 upon illumination by a UV lamp (λ_{ex}=375 nm). The concentration of C343 in solution compared to the ADB linker from left to right is 0.005, 0.1, 1, 10, and 30%. This figure is copied with permission from Oldenburg et al. [92]

The verification of the previously described procedure was done by the preparation of three new samples during summer and winter at each concentration of 2%, 1% and 0.1%. It was found that the maximum is always around 1% of C343 (shown in Figure 4.12). [92]

As expected the samples with C334 are not showing remarkable enhancement of the PLQY. Figure 4.13 shows the comparison of the PLQYs of ADB SURMOFs loaded with C343 and C334. [92]

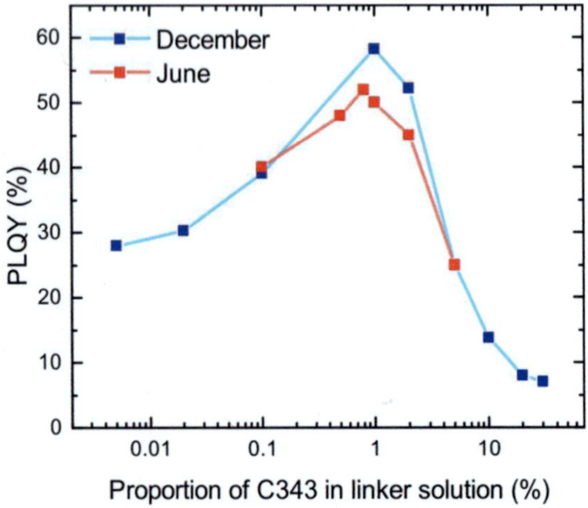

Figure 4.12: The PLQY of ADB SURMOFs with varying proportions of coumarin 343 in the linker solution are shown for fabrication runs made in December and June in Karlsruhe, Germany. This figure is copied with permission from Oldenburg et al. [92]

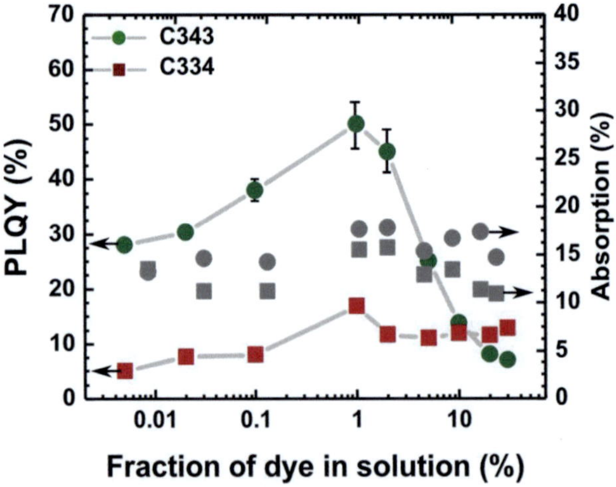

Figure 4.13: The graph shows the PLQY and absorption measurement of ADB SURMOFs infiltrated by C343 and C334 as a function of the concentration of the dye in the ADB linker solution. Samples with 2%, 1% and 0.1% of C343 are repeated 3 times for an error estimation that is plotted by an error bar. This figure is copied with permission from Oldenburg et al. [92]

4.4 Singlet exciton diffusion in ADB SURMOFs

The earlier experiments show that the effect of low C343 concentrations is still observable. This might be because the excitons are moving very fast within the ADB SURMOF. From the concentration of C343 within the ADB SURMOF a first rough estimation of the diffusion length can be made. This assumes that exciton diffusion occurs mostly into [010] direction, where the spacing between adjacent sheets is only 0.6 nm. Further, it is assumed that at 1% molar ratio the excited states are all reaching the C343 molecules. Related results were obtained for a truxene based MOF that was infiltrated with coumarin dyes as well. Also here the optimal doping level was 1% at which 90% of the excited states were quenched by the coumarin dyes. [101] The strong quenching in this SURMOF fabricated with the 1% coumarin solution can be used to estimate the diffusion lengths of excitons in the ADB SURMOF. From the XRD analysis, the closest distance between two linkers in an ADB SURMOF is 0.6 nm and if the C343 stays at a 1% dilution as it is in the solution, it can be assumed that two C343 molecules are separated by 60 nm. From this a diffusion length of 30 nm can be estimated, that is half the distance of two C343 molecules. Yet, the spatial concentrations of linkers in the SURMOF might differ from the solution. A better estimation can be obtained from the comparison of the ratio between the ADB (at 360 nm) and the C343 (at 450 nm) absorption peak (see Figure 4.14A) and comparing this absorption ratio to a solution with known molar ratio between ADB linkers and C343 molecules (see Figure 4.14B). By evaluating this, a curve could be extracted that shows the uptake of C343 in the SURMOF depending on the molar ratio in the linker solution (see Figure 4.14C). The behavior is linear for low concentrations of C343, however, becomes non-linear at high concentrations. This is due to the limited space within the SURMOF.

The provided analysis increases the separation between two C343 molecules to 100 nm. From both analysis a rough estimation of the diffusion length of 30 to 50 nm could be found. In comparison to other organic materials, the diffusion length is exceptional long,[24, 102, 103] yet comparable to singlet diffusion lengths reported for dyes ordered in nanochannels,[104] and for anthracene crystals.[105] Furthermore, the experiments agree well with the singlet exciton diffusion lengths of 48 nm in the truxene based MOF,[101] and the diffusion length of excitons of roughly 52 nm in a porphyrin-based MOF.[92, 106]

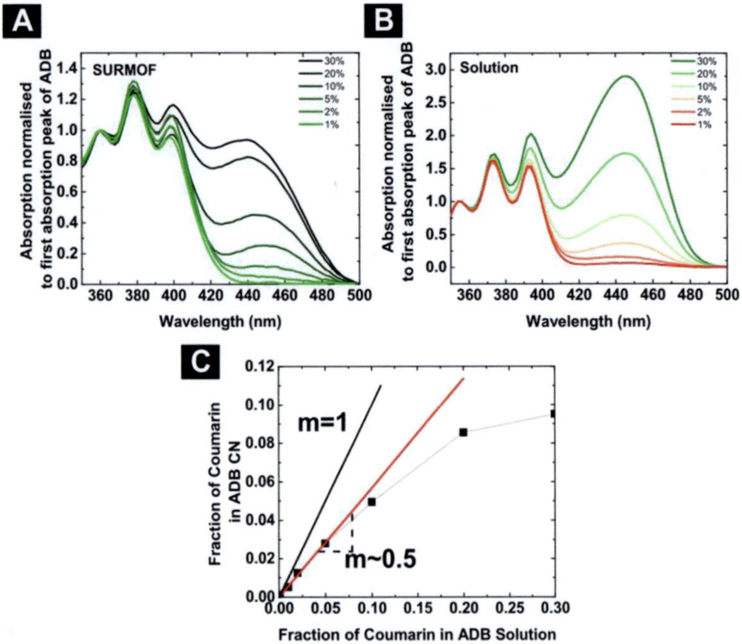

Figure 4.14: A) Absorption spectra of ADB SURMOF with different C343 molar ratios in the linker solution. B) Absorption spectra of ADB solution with different C343 molar ratios. C) Resulting correlation between the concentration of C343 in the SURMOF and the initial concentration. The red line shows a linear dependence for low concentrations with a slope of 0.5. As a reference, a linear function with the slope of 1 is plotted. This figure is copied with permission from Oldenburg et al. [92]

To learn more on the dynamics of the excited states in the ADB SURMOF, an intensity dependent PL analysis was performed. This was done in a pure ADB SURMOF to avoid any influence of the C343 molecules. For this a time-integrated PL emission was recorded after the excitation with a femtosecond pulse at 400 nm. From a non-linear behavior and a suggested rate equation model, the annihilation between the singlet excitons can be extracted. Figure 4.15A shows a blue shift in the emission spectrum from which can be concluded that the monomer-like states are annihilating not as efficient as the aggregate states, that is also due to the shorter lifetime. For a better analysis, the time-integrated PL intensity as a function of excitation intensity is plotted for both emission regions (see Figure 4.15B). The emission from the monomer-like excitons (integrated from 440–460 nm) and the aggregate excitons (emission integrated from 550-750 nm) show a non-linear dependence on the excitation fluence. [92] Which means that they are annihilating during their existence. By using the following rate equation for the excitons:

$$\frac{\partial A}{\partial t} = -kA - \gamma A^2, \tag{4.4}$$

where A is the excited-state density, k the monomolecular recombination constant and γ the bimolecular recombination constant the annihilation rate can be extracted. In the case of a three-dimensional diffusion, γ can be assumed to be time-independent. With $A(t = 0) = A_0$ as the excited state concentration directly after the excitation, the following solution can be found for above rate equation:

$$A_{3D}(t) = \frac{kA_0}{\exp(kt)\,(k + \gamma A_0) - \gamma A_0}. \tag{4.5}$$

The integration over time leads to the following expression: [92]

$$\int_0^\infty kA_{3D}(t)dt = \frac{k}{\gamma}\ln\left(1 + \frac{\gamma}{k}A_0\right). \tag{4.6}$$

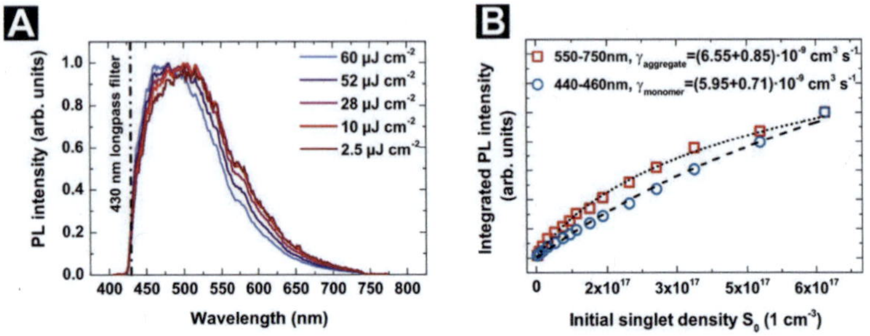

Figure 4.15: A) Normalized time-integrated emission spectra of a pure ADB SURMOF excited with different pulse energy densities. B) Spectrally integrated PL intensities (normalized to maximum) as functions of the initial singlet density S_0 This figure is copied with permission from Oldenburg et al. [92]

The initial excited state concentration can be calculated from the pump fluence, the fraction of photons absorbed (7%), and the ADB SURMOF thickness (140 nm). From the fit the parameter γ/k can be found, as it is the only fitting parameter. By fitting it was found, for the aggregate state $\gamma/k = (2.16 \pm 0.14) \cdot 10^{-17}\ \mathrm{cm}^3$ and for the monomer state $\gamma/k = (0.22 \pm 0.05) \cdot 10^{-17}\ \mathrm{cm}^3$. [92] From the lifetime obtained earlier (Figure 4.4B and C) for the monomer (0.37 ns) and for the aggregate emission (3.3 ns), the bimolecular recombination rates are $\gamma_{aggregate} = (6.55 \pm 0.85) \cdot 10^{-9}\ \mathrm{cm}^3\ \mathrm{s}^{-1}$ and $\gamma_{monomer} = (5.95 \pm 0.71) \cdot 10^{-9}\ \mathrm{cm}^3\ \mathrm{s}^{-1}$. The bimolecular recombination rate is related to the diffusion length $L_{3D} = \sqrt{6D/k}$ (assuming three-dimensional diffusion) [103] by:

$$\gamma = \frac{4}{3}\pi k R_0 L_{3D}^2, \tag{4.7}$$

where R_0 is the effective interaction radius of the annihilation. From this the parameter $R_0 L_{3D}^2$ can be extracted for the aggregate state $R_0 L_{3D}^2 = 5200 \pm 300\ \mathrm{nm}^3$ and for the monomer state $R_0 L_{3D}^2 = 525 \pm 120\ \mathrm{nm}^3$. Further, for the aggregate state an interaction radius is estimated to be at least 2 nm, resulting in a diffusion length of $L_{3D}^{aggregate} = 51 \pm 2\ \mathrm{nm}$.

Similar, for the monomer state an interaction radius of R_0~1 nm (one single ADB molecule) was assumed. The corresponding diffusion length is then $L_{3D}^{monomer} = 23 \pm 3$ nm for the monomer state. [92] The so found diffusion lengths agree well with the earlier estimations from the PLQY measurements. Yet, the SURMOF structure suggest that the diffusion into [010] direction is more likely, reducing the transport only to one dimension. [93] In the case of one-dimensional transport the annihilation constant γ is time dependent:[107]

$$\gamma = 4\pi D R_0 \frac{R_0}{\sqrt{2\pi D t}}. \tag{4.8}$$

With the time dependent annihilation constant, the solution for the rate equation in one dimensions is:

$$A_{1D}(t) = \frac{A_0 \exp(-kt)}{2\pi L_{1D} R_0^2 A_0 \operatorname{erf}(\sqrt{kt}) + 1}, \tag{4.9}$$

with $\operatorname{erf}(x)$ being the error function and $L_{1D} = \sqrt{2D/k}$ the one-dimensional diffusion length. An analytic expression for the integral $\int_0^\infty kA_{1D}(t)dt$ does not exist, and therefore the integral has to be calculated numerically and input into a non-linear regression method to fit the parameter $L_{1D}R_0^2$ (k is kept constant at the known value determined by the inverse lifetime). The resulting parameters are $L_{1D}R_0^2 = 655 \pm 20\ \text{nm}^3$ for the aggregate state and $L_{1D}R_0^2 = 145 \pm 23\ \text{nm}^3$ for the monomer state. As expected, the length for one-dimensional diffusion depends more strongly on the interaction radius.[24, 107] Assuming interaction radii (R_0) of 1, 2, 3 and 4 nm would result in diffusion lengths $L_{1D}^{aggregate}$ of 655, 163, 72, and 41 nm respectively for the aggregate state. For the monomer state the corresponding diffusion lengths $L_{1D}^{monomer}$ are 145, 36, 16 and 9 nm. [92]

4.5 Summary

The chapter described analyzed the emitter SURMOF based on the ADB linker. It was found that the PLQY in a pure ADB SURMOF is around 1%, meaning that the UCQY in a realized bilayer would remain low as well. Therefore, two strategies were analyzed to enhance the PLQY. First an optically inert linker (TPC) was used to dilute the ADB SURMOF. And indeed, this helped to increase the internal PLQY of the SURMOF. However, the external PLQY remained unchanged and more importantly, the transport properties cannot be remained with such a strategy. This is because two ADB linkers are not near each other anymore. The second strategy involved the infiltration of single dye molecules into the ADB SURMOF, namely C334 and C343. It was found that a carboxylic terminated molecule will stick better into the ADB SURMOF and a controlled variation of the concentration within the SURMOF could be shown for C343.

By varying the concentration of the C343 molecule the PLQY of the SURMOF could be optimized to nearly 52%, approach the 63% of the C343 molecule in solution. From this analysis a diffusion length between 30-50 nm was estimated and further verified by intensity dependent measurements.

A not included analysis is the question whether, the energy transport is FRET or DET. It might be that both are dominant in such a structure, where the distance between the molecules is very small. Further, a clear statement whether the transport is one-dimensional or three-dimensional could not be made. Faster time-resolved measurements could unveil the behavior.

4.5.1 Contributions to this chapter

Oldenburg, M. (Author):

- Idea on infiltration of C334 and C343 into ADB SURMOF during spraying and not after spraying.
- Synthesis of all ADB SURMOFs, ADB/TPC mixed SURMOFs, C343@ADB SURMOFs and C334@ADB SURMOFs.
- All structure characterization measurements (based on experimental setups that are built and maintained by other persons, see below) and analysis of these.
- All spectroscopic measurements (based on experimental setups that are built and maintained by other persons, see below) and analysis of these.
- Design of Figures and presentation of data.

Wollgarten, S.:

- Idea of infiltration of SURMOFs with molecules having carboxylic acid groups. This was demonstrated in Wollgarten's thesis. [17]

Turshatov, A.:

- Idea of creating a diluted SURMOF by using an optically inert linker (in this case TPC).

Busko, D.:

- Experimental setup for PLQY measurements.

Jakoby, M.:

- Discussions on the results about exciton annihilation and streak camera experimental setup.

Haldar, R. and Wöll, C.:

- Infrared spectroscopy measurements of C343@ADB SURMOFs.

Chen, K. and Hodgkiss, J. M.:

- Experimental setup for intensity dependent PL measurements.

Howard, I. A. and Richards, B. S.:

Guidance and scientific discussions and proofreading of the text.

5 Sensitizer layer, Pd(II) porphyrin based SURMOF

This chapter is based on parts of this publication:

M. Oldenburg, A. Turshatov, D. Busko, S. Wollgarten, M. Adams, N. Baroni, A. Welle, E. Redel, C. Wöll, B. S. Richards and I.A. Howard, *Photon Upconversion at Crystalline Organic–Organic Heterojunctions.* Adv. Mater., 2016, 28: 8477–8482.

This chapter analyses the properties of a Pd(II) porphyrin-based SURMOF. Porphyrin molecules that contain heavy atoms are used in TTA-UC system frequently due to their efficient ISC and long-lived triplet states. In this part of this work, the transition from Pd(II) porphyrin molecules in solution to the arrangement in a SURMOF structure is analyzed by steady-state and time-resolved spectroscopy measurements. From the findings, an estimation of the performance as a sensitizer layer in a bilayer approach is provided.

5.1 Structure analysis

The sensitizer SURMOF is based on the ditopic Pd(II) 5,15-diphenyl-10,20-di(4-carboxyphenyl) porphyrin (Pd-DCP) linker. For the coordination of Pd-DCP zinc metal centers are used. Figure 5.1A shows the schematic of the resulting SURMOF. The distance into [001] direction is estimated to be ~2.4 nm and into [010] direction it is ~0.6 nm. The out-of-plane XRD is shown in Figure 5.1B. Beside the main peaks that are representing the (001) diffraction, there is an additional peak between 4° and 5°. That means that the SURMOF is not monolithic but has crystals with other orientations as well.

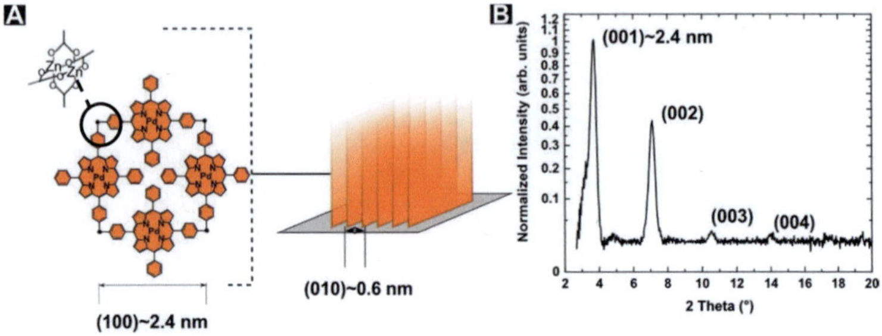

Figure 5.1: A) Schematic of the Pd-DCP SURMOF structure together with a zinc-based paddle wheel. B) The out-of-plane XRD of a Pd-DCP SURMOF. This figure is adapted with permission from Oldenburg et al. [92]

Figure 5.2 shows the FTIR absorbance of a Pd-DCP SURMOF on silicon. The bands at 1585 cm^{-1} and 1405 cm^{-1} are assigned to the asymmetric and symmetric stretching of COO^{-}. That indicates the formation of a zinc paddle-wheel structure with bidentate ligands as the difference Δ of both bands is between 140 and 200 cm^{-1}, namely 180 cm^{-1}. The formation of a paddle wheel structure is a strong indicator that the Pd-DCP SURMOF has no free carboxylic groups or unreacted zinc acetate moieties. [31, 92]

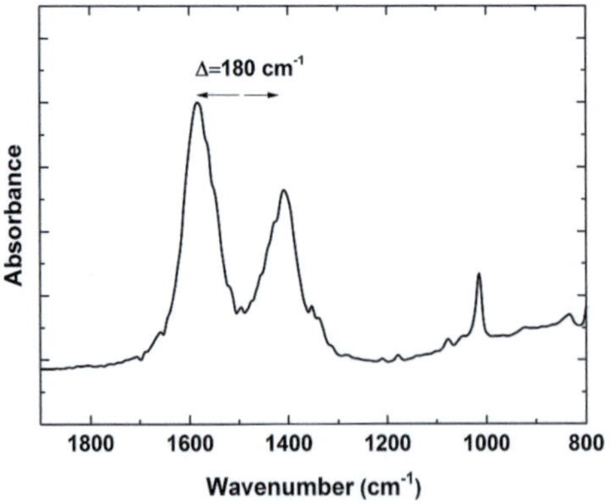

Figure 5.2: FTIR absorbance spectrum of a Pd-DCP SURMOF on silicon. The band difference of **Δ**=180 cm^{-1} indicates the formation of paddle wheel structures.

Figure 5.3 shows the morphological difference between three differently made Pd-DCP SURMOFs. The spray coated SURMOF that is shown in Figure 5.3A has a rough but dense surface. Figure 5.3B shows the spray coated Pd-DCP SURMOF after additional ultrasonication in an ethanol bath. The ultrasonication creates holes in the SURMOF (black area) and the remaining structure is as rough as before ultrasonication. In contrast to the spray coated Pd-DCP SURMOF, the surface of the spin-coated SURMOF shows a very flat morphology as Figure 5.3B demonstrates. [92, 108]

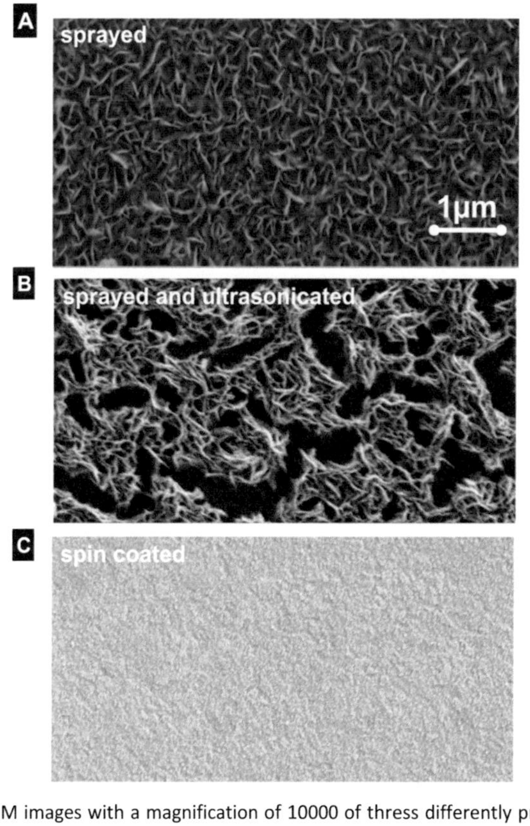

Figure 5.3: SEM images with a magnification of 10000 of thress differently prepared Pd-DCP SURMOFs. A) Spray coated Pd-DCP SURMOF. B) Spray coated and ultrasonicated Pd-DCP SURMOF. C) Spin-coated Pd-DCP SURMOF.

5.2 Performance as a sensitizer material

The close alignment of Pd-DCP molecules in a SURMOF leads to strong excitonic coupling. This is shown in the absorbance spectrum in Figure 5.4A. The Pd-DCP solution shows a strong Soret band at ~410 nm and a Q band absorbance at ~520 nm. The Soret band splits into to peaks in a SURMOF structure. De Miguel *et al* have shown that a Soret band splitting is due to densely packed porphyrin molecules. [109] That is indeed the case in a Pd-DCP SURMOF as the interplane distance is only ~0.6 nm. Further, the Q band absorbance shifts from ~522 nm to ~532 nm in the SURMOF. Both, the splitting of the Soret band and the red shift of the Q band indicate the formation of J and H aggregates in the Pd-DCP SURMOF.

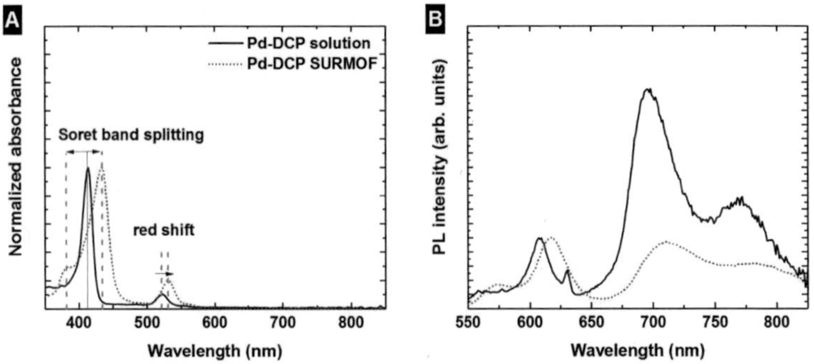

Figure 5.4: A) Absorbance spectrum of Pd-DCP solution (solid line) and SURMOF (dotted line). The dashed lines are guide to the eye showing the Soret band splitting and the red shift of the Q band. B) The PL emission spectrum of Pd-DCP solution (solid line) and SURMOF (dotted line). This figure is adapted with permission from Oldenburg et al. [108]

Figure 5.4B shows the emission spectrum of a Pd-DCP ethanol solution excited with a 532 nm cw laser. In comparison to this, the emission from a Pd-DCP SURMOF is red shifted by ~10 nm, which agrees well with the observation made in the absorbance spectrum. Further, the ratio between the peaks in solution and SURMOF is changed.

The performance of a sensitizer molecule for TTA-UC can be condensed from the two figures of merit, the UC-PLQY and the upconversion threshold. The UC-PLQY is the multiplication of the efficiencies of every intermediate state:

$$\phi_{UC} = \frac{1}{2} f \phi_{ISC} \phi_{ET} \phi_{TTA} \phi_{PLQY}. \tag{5.1}$$

The upconversion threshold is:

$$I_{TH} = \frac{1}{\alpha} \frac{1}{\phi_{ET}} \frac{{k_E^D}^2}{\gamma}. \tag{5.2}$$

The ISC efficiency ϕ_{ISC} and the energy transfer efficiency ϕ_{ET} can be limited by the sensitizer molecule. The ISC efficiency ϕ_{ISC} corresponds to the portion of singlet excitons that transform into triplet excitons. This can be estimated from the lifetime quenching of the fluorescence of a free-base porphyrin and the corresponding porphyrin with a heavy atom. Figure 5.5A shows the lifetime of the 620 nm peak of Pd-DCP SURMOF. The small difference between the system response function (solid line) and the PL signal demonstrates, that the lifetime is very short and has been estimated to be lower than ~10 ps.

This very short lifetime is evidence for strongly quenched fluorescence due to ISC. The created triplets inside the SURMOF are emitting light at ~710 nm. The transient shown in Figure 5.5B is composed of a fast and a slow component. The expectation lifetime that is calculated from a double exponential fit is ~300 μs. From the time-resolved data an ISC efficiency close to unity can be estimated. Further, the triplet lifetime is long allowing efficient energy transfer to an emitter molecule. [108]

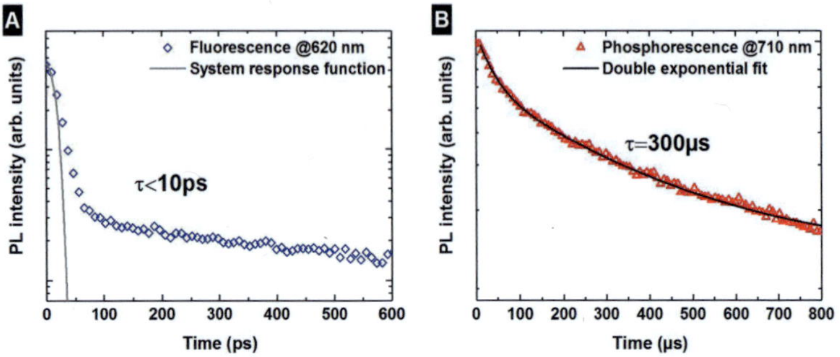

Figure 5.5: A) Transient at ~620 nm from Pd-DCP SURMOF (data points) and the system response function (solid line). The expectation lifetime is estimated to be shorter than 10 ps. B) Transient at ~710 nm (data points) and a double exponential fit (solid line). Additionally, the expectation lifetime of 300μs is indicated. This figure is adapted with permission from Oldenburg et al. [108]

5.3 Summary

The successfully created Pd-DCP SURMOF has been analyzed structurally and spectroscopically. It was found that the structure corresponds to the expected structure, derived from the length of the used linker. The spectroscopic analysis shows that the Pd-DCP SURMOF has a long living phosphorescence, that is necessary for efficient upconversion. Yet, the lifetime measurement shows that non-linear effects like TTA can happen in the Pd-DCP SURMOF as well. This means that the good transport properties might lead to loss mechanisms that will reduce the overall performance of UC in a bilayer. With this and the preceding chapter both types of SURMOFs have been synthesized and analyzed. The next chapter deals with multilayer SURMOFs that are made from Pd-DCP and ADB SURMOF.

5.3.1 Contributions to this chapter

Oldenburg, M. (Author):

- Synthesis of the Pd-DCP SURMOF and spectroscopic and structure analysis.
- Design of the figures and main text.

Dottermusch, S.:

- SEM measurements of the Pd-DCP SURMOF surface.

Heisler, S.:

- Infrared spectroscopy measurements.

Howard, I. A. and Richards, B. S.:

Discussion of the results.

6 Upconversion at SURMOF/SURMOF heterostructure interfaces

This chapter is based on parts of the following two publications:

M. Oldenburg, A. Turshatov, D. Busko, M. Jakoby, R. Haldar, K. Chen, G. Emandi, M. O. Senge, C. Wöll, J. M. Hodgkiss, B. S. Richards and I. A. Howard, *Enhancing the photoluminescence of surface anchored metal–organic frameworks: mixed linkers and efficient acceptors,* Phys. Chem. Chem. Phys., 2018, DOI: 10.1039/c7cp08452h.

M. Oldenburg, A. Turshatov, D. Busko, S. Wollgarten, M. Adams, N. Baroni, A. Welle, E. Redel, C. Wöll, B. S. Richards and I.A. Howard, *Photon Upconversion at Crystalline Organic–Organic Heterojunctions.* Adv. Mater., 2016, 28: 8477–8482.

In the preceding chapters, the emitter layer and the sensitizer layer are discussed isolated. In this final chapter, both layers will encounter each other. In a first approach, the ADB SURMOF is sandwiched between two Pd-DCP SURMOFs. The intention is to collect triplet excited states by Pd-DCP and concentrate them within the ADB layer. In the subsequent section, a more straightforward structure made only from two layers is created. The upconversion threshold is analyzed in both approaches and the dependence on the thickness is investigated. The second figure of merit, however, cannot be measured in these samples. Which is mainly due to the low PLQY of a pure ADB SURMOF. Therefore, the ADB SURMOF is loaded with C343. This increases the PLQY of the ADB SURMOF as already demonstrated. In the last section, the limitation by back-transfer is analyzed.

6.1 Trilayers

The overall efficiency of upconversion depends on the efficiencies of a cascade of events: ISC in the sensitizer, transfer of the triplet from sensitizer to emitter, creation of a singlet through triplet-triplet annihilation in the emitter, and then emission into the environment from this singlet state.

In Figure 6.1 a trilayer structure made from ADB and PdDCP SURMOFs is demonstrated. The three layer structure is prepared by the deposition of an ADB SURMOF layer on top of the Pd-DCP SURMOF layer. Subsequently, another Pd-DCP layer is deposited to create a thee layer structure that guarantees that excitons that are created in Pd-DCP can either decay or cross the interface between both SURMOFs. [108]

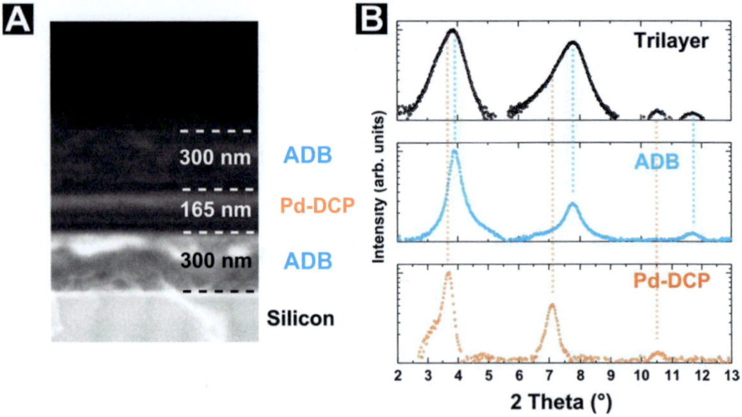

Figure 6.1: A) SEM cross-section of a three layer emitter-sensitizer-emitter (A-B-A) SURMOF heterostructure on a Si substrate. B) Out-of-plane XRD diffractograms that show the first three orders of the diffraction peak in the (001) crystalline plane of SURMOFs prepared from ADB and Pd-DCP alone, plus the trilayer. This figure is adapted with permission from Oldenburg et al. [108]

The cross-section of the trilayer is shown in the SEM image in Figure 6.1A. In the cross-section three different layers can be found. In this case, the thickness of the Pd-DCP SURMOF is determined to be 165 nm and the ADB SURMOFs to be around 300 nm. From XRD shown in Figure 6.1B the position of the diffraction peaks and the Bragg condition unit cell sizes for each SURMOF can be calculated. These are 2.2 nm and 2.4 nm for ADB and Pd-DCP respectively and coincide very well with similar structures. [31, 39] It is expected that such a small lattice mismatch between both SURMOFS won't lead to a large amount of defects. [44]

The diffractogram of the trilayer can be expressed as a sum of the single ADB and Pd-DCP layer. This shows that each layer is still crystalline and orientated in the [001]-axis perpendicular to the substrate. Additionally, to the XRD a ToF-SIMS measurement was performed. Figure 6.2 shows the depth profile of a trilayer. The Pd signal shows as expected a hump in the center, whereas the Zn signal drops. The Si signal was used to substrate. The smeared-out profiles show that the SURMOF is a rather soft material.

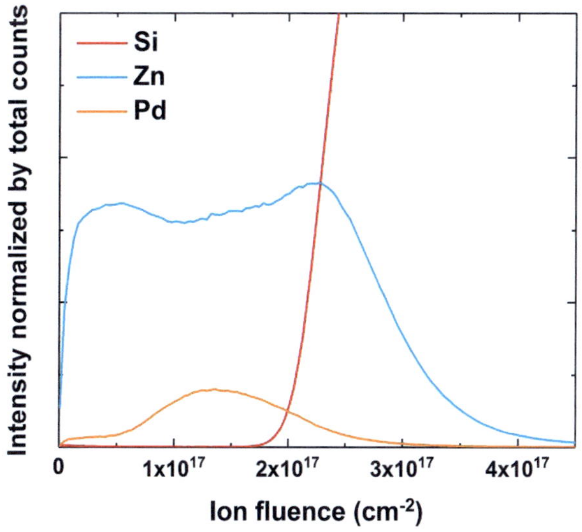

Figure 6.2: ToF-SIMS profiles from the trilayer SURMOF structure.

The SEM, XRD and ToF-SIMS characterization show that the trilayer is of sufficient quality. [108]

The mechanism of TTA-UC between the two SURMOFs is illustrated in Figure 6.3A. We expect that the diffusion of excitons from Pd-DCP is good enough so that a crossing to the ADB SURMOF is possible. In the ADB SURMOF both triplets annihilate and create a blue photon.

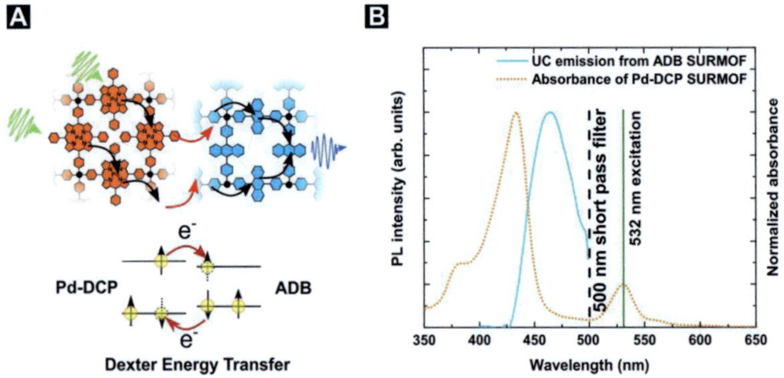

Figure 6.3: A) Schematic diagram of TTA-UC wherein two 532 nm photons absorbed by the sensitizer Pd-DCP SURMOF create triplet states that upon reaching the interface can transfer to the emitter ADB SURMOF by a Dexter two electron exchange mechanism. When two triplets meet in the emitter layer they can annihilate and emit a single higher energy photon. B) The absorption spectrum of Pd-DCP layer is shown (dotted line) and the wavelength of the excitation laser is shown as a vertical line at 532nm. The observed upconverted emission is shown in blue. This figure is adapted with permission from Oldenburg et al. [108]

The excitons diffuse via Dexter electron exchange mechanism, that relies on the transfer of two electrons (see Figure 6.3A). Thus, the observation of UC emission from the trilayer as shown in Figure 6.3B is evidence that electrons can move across the interface. [108] As discussed previously, the layer approach should present an opportunity to lower the upconversion threshold based on controlling the sensitizer and emitter layer thicknesses. Therefore, several trilayer structures were created and the intensity dependence measured (see Figure 6.4A). Firstly, it was necessary to proof that with thicker getting Pd-DCP layers the threshold drops. Yet, the threshold analysis for different Pd-DCP layers shown in Figure 6.4B shows the opposite behavior. Namely, that the thinnest Pd-DCP layer has the smallest threshold: 25 ± 4 mW cm^{-2}. The thicker Pd-DCP layer thresholds are: 45 ± 2 mW cm^{-2}, 73 ± 2 mW cm^{-2}, and 117 ± 2 mW cm^{-2} at layer thicknesses of 32 nm, 64 nm, and 165 nm, respectively. [108]

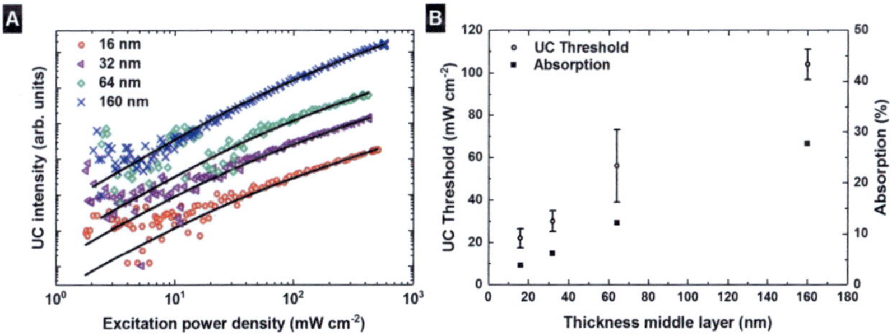

Figure 6.4: A) Upconversion optical emission intensity as a function of 532 nm optical excitation intensity for A-B-A samples with different sensitizer layer thicknesses. The threshold is determined by a fit to the predicted intensity dependence curve. B) The upconversion thresholds as a function of sensitizer layer thickness. This figure is adapted with permission from Oldenburg et al. [108]

In the theoretical discussion about bilayers one assumption was that the diffusion length of the excitons is larger than the layer thicknesses of sensitizer or emitter. Additionally, it was assumed by proper boundary conditions that the transfer happens instantaneously. Hence, one of these assumptions does not hold for the created trilayers. From earlier measurements on the ADB SURMOF it could be demonstrated that the diffusion lengths in SURMOFs are remarkably long. Thus, it can be assumed that the limiting process is the transfer across the interface. To improve this ultrasonication or spin-coating can be used. [92, 108]

6.2 Bilayers

The next samples that were created were only bilayers. This increases the concentration of sensitized triplets in the ADB layer, that should lower the threshold. Additionally, between the deposition of ADB on Pd-DCP SURMOF an ultrasonication step was included. 9 different samples, with 3 different layer thicknesses for ADB and Pd-DCP SURMOFs were created and the threshold was analyzed. Remarkably they show a lot lower threshold than the trilayer samples. Figure 6.5A shows the intensity dependence measurement of the sample with the lowest threshold achieved, it is less than 1 mW cm^{-2}. Figure 6.5B shows that the threshold depends as expected on the thickness od Pd-DCP and ADB layer. [92, 108]

The sonication step seemed to improve the interface quality so that the created triplets can safely pass the valley between Pd-DCP and ADB SURMOF. A timeresolved analysis of the phosphorescence gives an estimate on how many excitons are quenched by the ADB layer (see Figure 6.5C). The curves for the neat Pd-DCP and the bilayer with the lowest threshold show a non monoexponential decay. Therefore, the triplet transfer efficiency was estimated by using the following equation: $\phi_{\mathrm{tr}}^{\mathrm{E}\rightarrow\mathrm{S}} = 1 - \frac{\int I_{BA}(t)dt}{\int I_B(t)dt} = 1 - \frac{(\alpha_1\tau_1+\alpha_2\tau_2)_{BA}}{(\alpha_1\tau_1+\alpha_2\tau_2)_B}$, where $\alpha_i\tau_i$ are the weights and lifetimes extracted from the biexponential fit for the Pd-DCP and the bilayers. From this fit a triplet transfer efficiency from Pd-DCP to ADB SURMOF of 58±6 % could be estimated. Except for bad transfer across the interface another reason that the triplets might not get transferred is that they do not reach the interface as the thickness of Pd-DCP is too large.

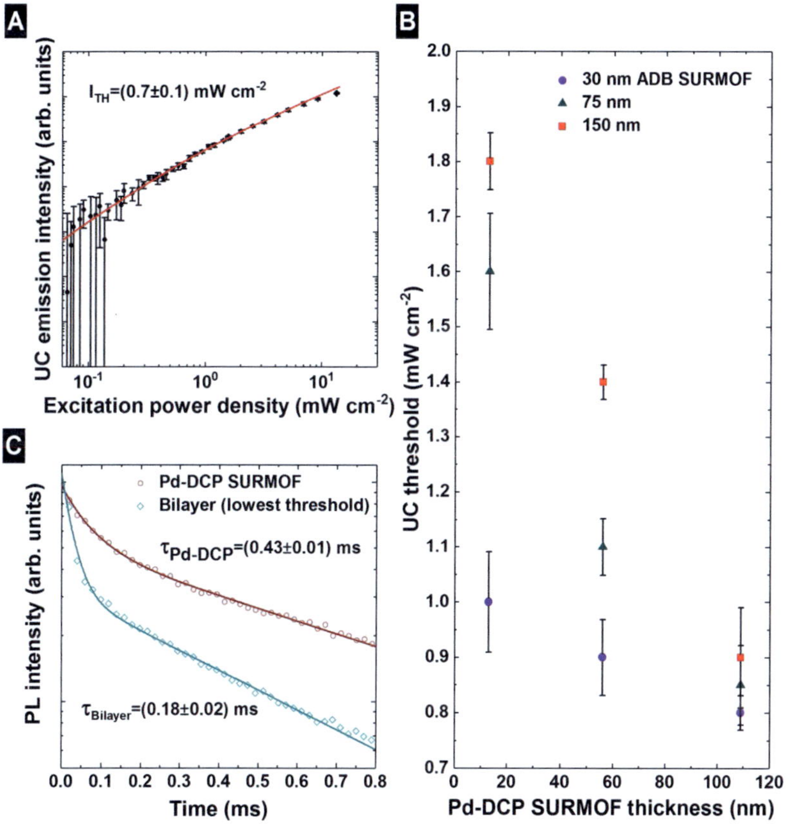

Figure 6.5: A) The threshold measurement for the best performing bilayer (Pd-DCP 110 nm, ADB 30 nm) B) The upconversion thresholds for all 9 B-A samples. C) Time-resolved phosphorescence measurements for a sensitizer single layer (110 nm thick) and the lowest threshold bilayer. This figure was adapted with permission from Oldenburg et al. [108]

Figure 6.6 shows the PL spectrum of a Pd-DCP SURMOF and two bilayers. The thickness of the ADB in the bilayers is kept constant, but the Pd-DCP is either 8 or 10 cycles thick (128 nm or 160 nm). The phosphorescence signal at ~700 nm is quenched in both bilayers. However, the quenching in the bilayer with the thinner Pd-DCP is stronger.

Although the threshold is low for the bilayers, the quantum efficiency is low as well. Looking again on the equation for the UC quantum yield: $\phi_{UCQY} = \frac{1}{2} f \phi_{PLQY} \phi_{TTA} \phi_{DET}^{E \to S} \phi_{ISC}$, the only two factors that might limit the quantum yield are f and the quantum yield of the emitter. Here f is the fraction of triplet-triplet collisions that generate a singlet, and is around 0.4. [63] ϕ_{PLQY} is the photoluminescence quantum efficiency of the emitter, that was measured to be less than 0.01. $\phi_{DET}^{E \to S}$, that has been found to be approximately 0.6. From these estimations the upconversion efficiency ϕ_{UCQY} for the demonstrated bilayers should be less than roughly 0.1%. [108] However, by using C343 the quantum yield of the emitter can be significantly in increased from under 1% to roughly 52%. This means, by using a C343@ADB SURMOF as an emitter the upconversion quantum yield should increase to 5.2%! Further, instead of using ultrasonication spin coating is used to guarantee a smooth interface. Whether, this strategy works is shown in the next section.

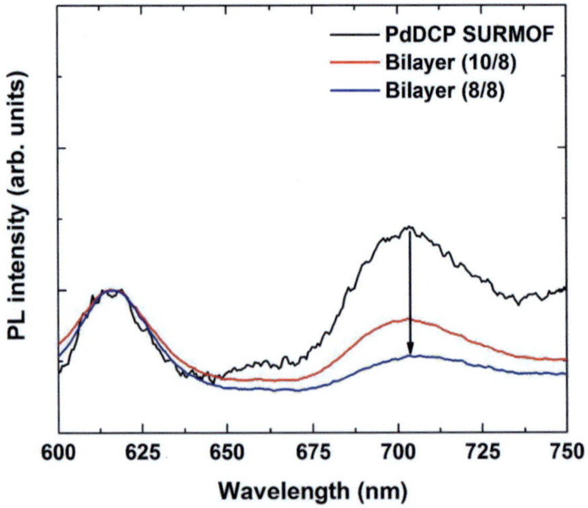

Figure 6.6: PL spectra for a Pd-DCP SURMOF and two bilayers in which the thickness of the ADB SURMOF is kept constant but the Pd-DCP is varied to be 10 or 8 cycles (160 or 128 nm) thick. The arrow indicates the quenching of the phosphorescence signal at ~700nm.

6.3 Enhanced Bilayers

Two bilayers were created. Both, the Pd-DCP and C343@ADB SURMOF were spin coated onto two glass substrates. The spin cycles for the Pd-DCP SURMOF were 10, meaning that the created sensitizer layers have a thickness of roughly 24 nm. The ADB SURMOF thickness was varied to be either 8 cycles (~18 nm) or 4 cycles (~9 nm).

Figure 6.7A shows XRD of the Pd-DCP SURMOF, the C343@ADB SURMOF and the Bilayer (8 cycles Pd-DCP). The diffractograms of each SURMOF structure coincide with the discussion provided for a trilayer structure in Figure 6.1B. [92]

Interestingly, the emission spectrum shown in Figure 6.7B for the upconversion emission (upon excitation with 532 nm) does not coincide with the direct excitation of a C343@ADB SURMOF. Instead, the emission spectrum agrees well with the emission peak of the monomer-like PL spectrum from a pure ADB SURMOF, that is shown as a hatched curve. However, if the C343@ADB SURMOF within a bilayer is directly excited (at 370 nm) the maximum of the PL spectrum shows no strong difference compared to single layer C343@ADB layer. This demonstrates that there is a clear difference between the emission spectrum of the C343@ADB (1%) bilayer that is excited by TTA compared to when it is excited directly. The reason could be that the excitons show a very anisotropic diffusion, meaning that they stay across the interface instead of moving from it away. This blocking of the excitons can further lead to backtransfer to the Pd-DCP SURMOF. [92] This effect of backtransfer is analyzed in the next section and will show the limitation of the current molecular architecture of SURMOFs.

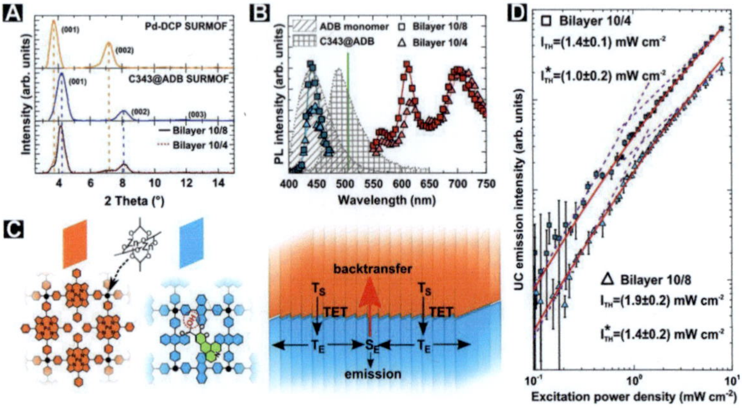

Figure 6.7: A) Coplanar XRD showing the crystallinity and orientation of the Pd-DCP SURMOF, the C343@ADB SURMOF and the bilayer made from C343@ADB SURMOF and 8 cycles of Pd-DCP SURMOF. The peaks are labeled with the orientation. Additionally, the dashed lines indicate the composition of the bilayer from the monolayers. B) PL emission spectrum of the two bilayers together with the emission spectra of ADB monomer emission and C343@ADB emission as filled curves. C) Schematic of the SURMOF heterostructure bilayers together with the flow of triplet excitons across the interface. D) Intensity dependence measurements for the UC emission of both bilayers together with the analytic fit (solid lines) and the approximated asymptotes (dashed lines) with a slope of two and one respectively. Additionally, thresholds from the analytic fit and the approximated asymptotes (indicated by an asterisk) are noted. This figure is copied with permission from Oldenburg et al. [92]

6.3.1 Back-transfer from ADB to PdDCP SURMOF

Direct evidence, that excitons are backtransferred to the Pd-DCP SURMOF when they are close to the interface was provided by timeresolved spectroscopy. The hypothesis is that excitons are travelling rather between the sheets (that are separated only by 0.6 nm) rather than along the sheets (that have a minimal separation of ~2 nm). In the case of a bilayer that performs TTA-UC, the created triplets are crossing the interface and are staying there as it is shown in Figure 6.7C. Now two things can happen, the triplets are annihilating and emitting a blue photon or are decaying towards aggregate like states. The created aggregated states have stronger overlap with the absorption band of the Pd-DCP SURMOF and so can be absorbed again. Further, the if the aggregate state gives its energy to a C343 molecule that is located close to the interface as well it will be absorbed by Pd-DCP again. Thus, the remaining excited state that can be sufficiently detected is the monomolecular-like state from ADB SURMOF. [92]

Figure 6.8A shows the two-dimensional streak data for a pure C343@ADB SURMOF and the two bilayers that were created earlier. Figure 6.8B shows the decay of the emission wavelengths from 470-510 nm (that corresponds to the PL emission peak coming from C343), the PL lifetime of the bilayer is shorter than that of the C343@ABD. This is evidence, that the excited-states on C343 molecules can give their energy back to the Pd-DCP SURMOF. PL from 410-450 nm that mainly comes from the monomer-like ADB states is not as strongly quenched; the lifetime in the bilayer is only slightly shorter than in the C343@ABD alone. These observations are consistent with the hypothesis that the upconverted emission comes from a small amount of emission from the monomer-like state before it can transfer to C343. Unfortunately, the C343 emission – that would be much stronger – is not observed due to the fast return of the excited-state on the C343 back to the sensitizer layer. To further support this hypothesis a mixed SURMOF is made. This SURMOF consisted from ADB, Pd-DCP and C343. The C343 is kept to be 1% in the ADB linker solution. The amount of Pd-DCP is varied to be 0%, 0.1%, 1% and 10% in the linker solution.

A scheme is shown in Figure 6.9A. The absorbance spectra from all four samples are shown in Figure 6.9B. Clear evidence of C343 comes from the peak at ~475 nm. Further, the Soret band from Pd-DCP starts to increase as the concentration of Pd-DCP in the linker solution is increased.

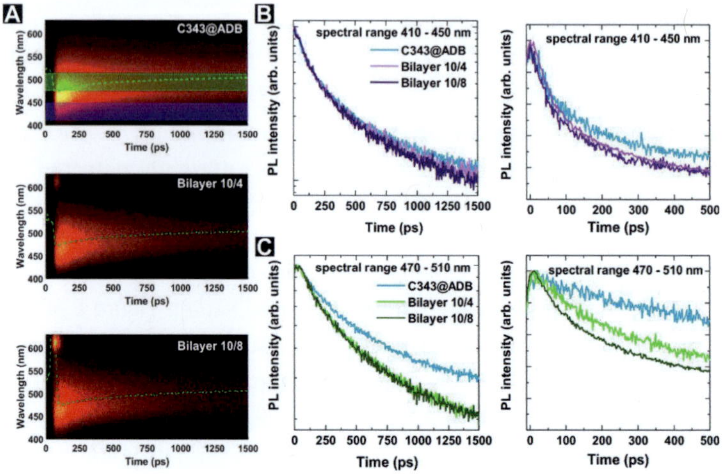

Figure 6.8: A) Streak camera measurements of the C343@ADB, bilayer 10/4 and bilayer 10/8. The green and violet areas indicate the spectral range, which are integrated over the wavelength to extract the corresponding transients. The green dashed line indicates the position of the maximum at a given time slice. B) Transients for the spectral range 410 – 450 nm. Left: Time range 0 – 1500 ps. Right: Time range 0 – 500 ps. The transients show that the dynamic quenching in the bilayers via FRET back transfer is small. C) Transients for the spectral range 470 – 410 nm. Left: Time range 0 – 1500 ps. Right: Time range 0 – 500 ps. The transients show that the dynamic quenching in the bilayers is significant for this spectral region. This figure was copied with permission from Oldenburg et al. [92]

Figure 6.9C shows the PL of all four samples with strong phosphorescence coming from the sample with the highest concentration of Pd-DCP. Interestingly, the PLQY decreases with increasing Pd-DCP concentration as is demonstrated in Figure 6.9D. Another evidence, that Pd-DCP is quenching the C343 PL is the time-resolved date in Figure 6.9E. The backtransfer is drastically reducing the upconversion quantum yield, however has no effect on the threshold. [63] Figure 6.7D shows the measured data for the intensity dependence of the upconversion. The threshold is obtained through direct fit of the theoretical intensity dependence curve and is compared to the alternative method of using the asympotetes. For both bilayers the threshold was below 2 mW cm^{-2}.[92]

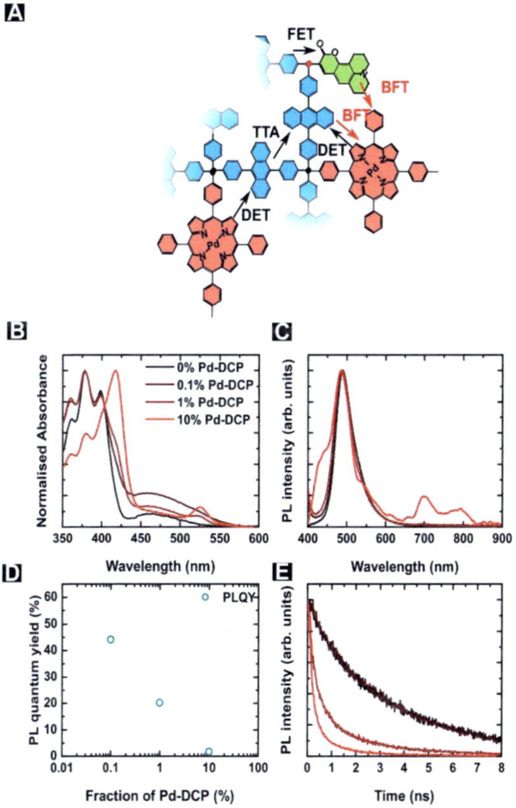

Figure 6.9: A) Schematical drawing of the mixed SURMOF made from ADB, C343 and Pd-DCP. The arrows illustrate possible energy path ways (DET – Dexter energy transfer, FET – Förster energy transfer, TTA – Triplet-triplet annihilation, BFT – back energy transfer). B) The Absorbance spectra of the mixed SURMOF with varying Pd-DCP concentration. C) The corresponindg PL spectra of the four mixed SUMROF samples. D) The PLQY of the three SURMOFs that contain Pd-DCP. D) The time-resolved PL for 400-600 nm.

6.4 Summary

In this chapter the TTA-UC performance was analyzed for a tri and a bilayer SURMOF structure. It was found that the smoothness across the interface is very important to achieve low thresholds. The found thresholds were <1 mW cm^{-2}. This is a significant achievement in the field of solid-state TTA-UC. However, using the pure ADB SURMOF as an emitter lead to a low UCQY. To increase this, the strategy by infiltrating C343 into the SURMOF was used. The so created enhanced bilayer show very low threshold, but again the UCQY was so low that it could not be measured. From time-resolved measurements it was concluded that the sensitizer acts as a strong energy acceptor, which means that even though there are upconverted singlet excitons many of them are back transferred to the sensitizer.

6.4.1 Contribution to this chapter

Oldenburg, M.:

- Synthesis and characterization (spectroscopic and structure) of all tri- and bilayer structures.

Emandi, G. and Senge, M. O.:

- Synthesis of the Pd-DCP linker.

Wollgarten, S.:

- Reproducibility check of ADB SURMOFs.

Adams, M. and Baroni, N.:

- Discussion of the obtained results.

Welle, A.:

- ToF-SIMS measurements.

Redel, E. and Wöll, C.:

Introduction to the spraying setup.

7 Conclusion and outlook

7.1 Conclusion

The physics of organic spectral upconversion are analyzed for two cases, an isotropic distribution of emitter and sensitizer molecules and a bilayer structure. The main findings of the former system are that there is an upconversion threshold I_{TH} that sets the value for the excitation power density from which upconversion emission approaches the maximum efficiency. The threshold is given by the following equation

$$I_{TH} = \frac{1}{\alpha} \frac{1}{\phi_{DET}} \frac{{k_E^D}^2}{\gamma}. \tag{7.1}$$

The main conclusion from this equation is that the absorption and energy transfer efficiency should be high, the emitter triplet lifetime long and the diffusivity of the emitter triplets high. The predicted intensity dependence could be verified accurately in an experiment using Pd-TBP as a sensitizer molecule and perylene as the emitter molecule. From the time dynamics, an estimate of a steady-state time is derived which is around the double of the lifetime of emitter triplets. The rate equations are further verified by time-resolved measurements. The excellent agreement between experimental data and theory is the basis for the analysis of a bilayer structure.

Within a bilayer system, the sensitizer and emitter are separated into two distinct layers. An important finding is that the emitter triplet concentration depends on the thickness of the sensitizer d_S and the thickness of the emitter d_E:

$$E_T \propto G \frac{d_S}{d_E}. \tag{7.2}$$

This behavior predicts that a thick sensitizer layer will transfer more triplets to the emitter layer than a thin layer. Furthermore, the concentration of emitter triplets will rise for thin layers. This geometric concentration of the emitter triplet concentration increases the probability of collisions. Effectively the threshold decreases in a bilayer approach. It must be stated, however, that this theoretical derivation assumed isotropic movement of the excitons, a large diffusion length and total quenching at the interface. Meeting all three requirements with one material is a very difficult task. SURMOFs seemed to fulfill at least the larger diffusion lengths. Isotropy of exciton transport and a high-quality interface between two different SURMOFs is not necessarily realized in SURMOFs and has to be kept in mind.

Two candidates were chosen to realize bilayers based on SURMOFs that perform TTA-UC. Pd-DCP as the sensitizer and ADB as the emitter. Although, these molecules work very well in solutions it is not obvious that they will work in a SURMOF structure. For this a more detailed analysis of the energy level has to be performed and maybe there are molecules that would fit better. Nonetheless, the ADB SURMOF has shown very interesting physics. From detailed spectroscopic analysis it was found that the energy transport is very efficient and fast. This is mostly due to the close proximity of ADB linkers, which however also introduces strong losses in form of aggregated states. However, also this could be solved by introducing bright molecules into the framework. C343 was the candidate that was chosen and that helped to increase the PLQY from 1% to 52%. Yet, there might be better candidates as energy valves. Together with Pd-DCP SURMOF, the ADB SURMOF was used to create first layered SURMOF structures that perform TTA-UC. It was found that the interface quality must be very good to observe the predicted behavior of the threshold on the thickness of the two layers. However, in both SURMOFs an isotropic energy transfer is not observed. This lead to strong backtransfer of the upconverted energy from the ADB SURMOF to the Pd-DCP SURMOF. The final bilayer samples have a very low threshold (<2 mW cm^{-2}), however, the quantum yield is still too low for any applications.

7.2 Outlook

Although the quantum yield of the so created SURMOF based TTA-UC heterostructures is low, the threshold is extremely low. This is making the created structures superior to any other previously developed TTA-UC solid-state systems. Based on the findings about the transport properties within the SURMOFs the author wants to make suggestions for further improvements based on his intuition.

From the geometry of the SURMOFs, an anisotropic transport of excitons is likely. The intrasheet distance of around 0.6 nm makes the excitons move better between the sheets rather than along the sheets. In a heterostructure based this means that the excited states will stay close to the interface making the back-transfer a main loss channel. An improvement would be to flip the orientation of the SURMOFs. This would increase the number of excited states that move away from the interface in a bilayer. However, here as well the highest possible amount of excited states that move away from the interface is 50%. Therefore, another layer between the sensitizer and emitter SURMOF has to be introduced, that ideally possess a triplet energy level that lies between the triplet level of the sensitizer and the emitter and a singlet level that lies well above the singlet level of the sensitizer and the emitter. Both improvements would lead to a better heterostructure for TTA-UC.

The flipping would also make other devices possible. OLEDs based on SURMOFs were built, however, the measurements were not successful and so are not included in this thesis. The conductivity of SURMOFs was found to be very low and based on this it was concluded that an OLED cannot work as not enough charge carriers could be injected. This might be as well due to the anisotropy of the SURMOFs. This means that injected charge carriers can only travel from sheet to sheet but not along the sheets keeping them from meeting and so emitting light.

Another tremendous important fact for any future devices based on SURMOFs is there salt-like nature. The organic linker and the metal center are bound by ionic forces, making a flow of charges impossible as all free-charges are kept at the metal center or the organic linker. However, it was shown that the metal center can be removed after the creation of a SURMOF. The remaining SURGEL has a partially intact SURMOF structure and might be a better candidate for devices like OLEDs or similar.

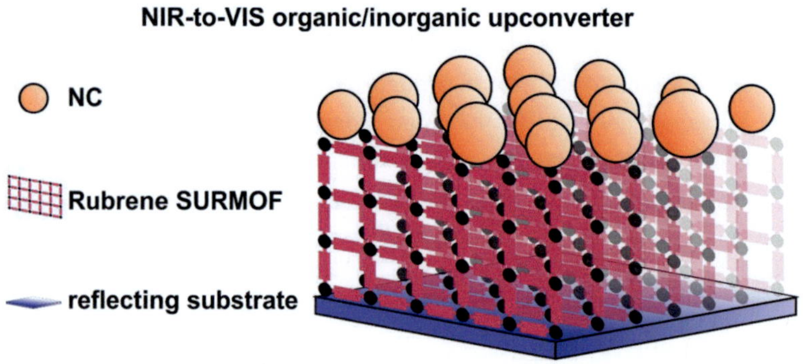

Figure 7.1: Concept of a NIR-to-VIS organic/inorganic upconverter based on a rubrene SURMOF as the emitter and PbS nanocrystals (NC). The Absorption of PbS NCs is higher then the bandgap of certain perovskite solar cells. With this it might be feasible to create a SURMOF that has a low threshold and the right wavelength shift.

Further, the synthesis procedure that is indeed improved by spin coating is still not well enough developed. A better control over parameters like humidity and temperature and substrate functionalization might play an important role. In conclusion the author thinks that SURMOFs in their current form are not capable of any electric performance and so is limited in the use as an active material in an OPV or OLED. However, for UC SURMOFs might play an important role. Figure 7.1 shows a concept for a future up-converter design based on a SURMOF. The SURMOF is made from rubrene, that is a prominent red emitter for upconversion. The sensitizer in this case are PbS nanocrystals (NC), that absorb in the infrared. The threshold will be low due to the SURMOF and the wavelengths are right for the use of the upconverter in perovskite solar cells. The synthesis route for a rubrene linker is shown in Figure 7.2.

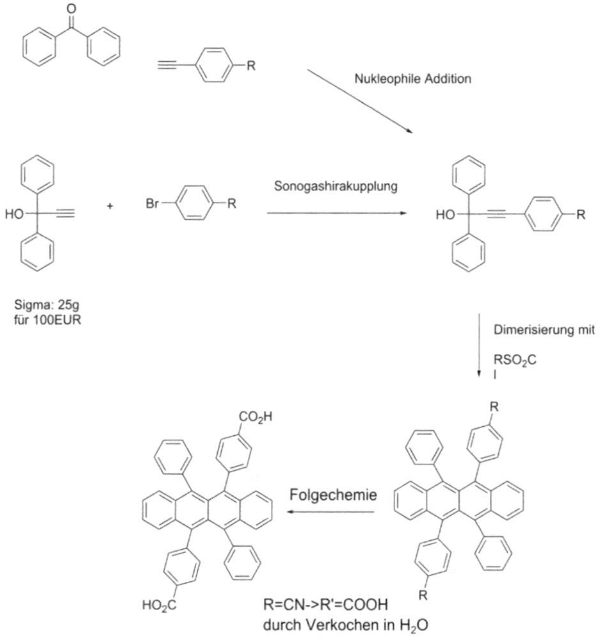

Figure 7.2: The synthesis scheme for a rubrene linker that could be used for a SURMOF.

Another approach is to coat the PbS NCs with a thin layer SURMOF and then infiltrate these UC nanoparticles into a perovskite.

References

1. Schulze, T.F. and T.W. Schmidt, Photochemical upconversion: present status and prospects for its application to solar energy conversion. Energy & Environmental Science, 2015. 8(1): p. 103-125.
2. Guter, W., et al., Current-matched triple-junction solar cell reaching 41.1% conversion efficiency under concentrated sunlight. Applied Physics Letters, 2009. 94(22): p. 223504.
3. Essig, S., et al., Raising the one-sun conversion efficiency of III–V/Si solar cells to 32.8% for two junctions and 35.9% for three junctions. Nature Energy, 2017. 2: p. 17144.
4. Bush, K.A., et al., 23.6%-efficient monolithic perovskite/silicon tandem solar cells with improved stability. Nature Energy, 2017. 2: p. 17009.
5. Yoshikawa, K., et al., Silicon heterojunction solar cell with interdigitated back contacts for a photoconversion efficiency over 26%. Nature Energy, 2017. 2: p. 17032.
6. Futscher, M.H. and B. Ehrler, Efficiency Limit of Perovskite/Si Tandem Solar Cells. ACS Energy Letters, 2016. 1(4): p. 863-868.
7. Kim, Y., et al., Solar-light photocatalytic disinfection using crystalline/amorphous low energy bandgap reduced TiO_2. Scientific Reports, 2016. 6: p. 25212.
8. Grove, W.R., LXXII. On a gaseous voltaic battery. The London, Edinburgh, and Dublin Philosophical Magazine and Journal of Science, 1842. 21(140): p. 417-420.
9. Eberle, U., B. Müller, and R. Helmolt, Fuel cell electric vehicles and hydrogen infrastructure: Status 2012. Vol. 5. 2012. 8790-8798.
10. Bloembergen, N., Solid State Infrared Quantum Counters. Physical Review Letters, 1959. 2(3): p. 84-85.
11. Auzel, F., Upconversion and Anti-Stokes Processes with f and d Ions in Solids. Chemical Reviews, 2004. 104(1): p. 139-174.
12. Goldschmidt, J.C. and S. Fischer, Upconversion for Photovoltaics – a Review of Materials, Devices and Concepts for Performance Enhancement. Advanced Optical Materials, 2015. 3(4): p. 510-535.

13. Hoseinkhani, S., et al., Achieving the photon up-conversion thermodynamic yield upper limit by sensitized triplet-triplet annihilation. Physical Chemistry Chemical Physics, 2015. 17(6): p. 4020-4024.
14. Kasha, M., Characterization of electronic transitions in complex molecules. Discussions of the Faraday Society, 1950. 9(0): p. 14-19.
15. McClure, D.S., Triplet-Singlet Transitions in Organic Molecules. Lifetime Measurements of the Triplet State. The Journal of Chemical Physics, 1949. 17(10): p. 905-913.
16. Marian, C.M., Spin-orbit coupling and intersystem crossing in molecules. Wiley Interdisciplinary Reviews: Computational Molecular Science, 2012. 2(2): p. 187-203.
17. Wollgarten, S., Triplet Dynamics for spectral upconversion using surface-bounded metal-organic-frameworks for organic dyes, in Department of Electrical Engineering and Information Technology. 2016, Karlsruhe Institute of Technology. p. 56.
18. Förster, T., Zwischenmolekulare Energiewanderung und Fluoreszenz. Annalen der Physik, 1948. 437(1-2): p. 55-75.
19. Yost, S.R., et al., Triplet vs Singlet Energy Transfer in Organic Semiconductors: The Tortoise and the Hare. The Journal of Physical Chemistry C, 2012. 116(33): p. 17369-17377.
20. Roy, R., S. Hohng, and T. Ha, A Practical Guide to Single Molecule FRET. Nature methods, 2008. 5(6): p. 507-516.
21. Dexter, D.L., A Theory of Sensitized Luminescence in Solids. The Journal of Chemical Physics, 1953. 21(5): p. 836.
22. Sudha Devi, L., et al., Triplet energy transfer in conjugated polymers. I. Experimental investigation of a weakly disordered compound. Physical Review B, 2008. 78(4).
23. Hoffmann, S.T., et al., Triplet energy transfer in conjugated polymers. III. An experimental assessment regarding the influence of disorder on polaronic transport. Phys. Rev. B, 2010. 81(16).
24. Tamai, Y., et al., One-Dimensional Singlet Exciton Diffusion in Poly(3-hexylthiophene) Crystalline Domains. The Journal of Physical Chemistry Letters, 2014. 5(2): p. 399-403.
25. Tamai, Y., et al., Exciton Diffusion in Conjugated Polymers: From Fundamental Understanding to Improvement in Photovoltaic Conversion Efficiency. J. Phys. Chem. Lett., 2015. 6(17): p. 3417-28.

26. Farha, O.K., et al., De novo synthesis of a metal–organic framework material featuring ultrahigh surface area and gas storage capacities. Nature Chemistry, 2010. 2(944-948).
27. Rodenas, T., et al., Metal–organic framework nanosheets in polymer composite materials for gas separation. Nature Materials, 2015. 14(1): p. 48-55.
28. Zheng, F., Y. Yang, and Q. Chen, High lithium anodic performance of highly nitrogen-doped porous carbon prepared from a metal-organic framework. Nature Communications, 2014. 5.
29. Talin, A.A., et al., Tunable Electrical Conductivity in Metal-Organic Framework Thin-Film Devices. Science, 2014. 343(6166): p. 66-69.
30. Erickson, K.J., et al., Thin Film Thermoelectric Metal-Organic Framework with High Seebeck Coefficient and Low Thermal Conductivity. Advanced Materials, 2015. 27(22): p. 3453-9.
31. Liu, J., et al., Photoinduced Charge-Carrier Generation in Epitaxial MOF Thin Films: High Efficiency as a Result of an Indirect Electronic Band Gap? Angew. Chem. Int. Ed., 2015. 54(25): p. 7441-5.
32. Deria, P., et al., Beyond post-synthesis modification: evolution of metal-organic frameworks via building block replacement. Chemical Society Reviews, 2014. 43(16): p. 5896-5912.
33. Lin, J., et al., Triplet Excitation Energy Dynamics in Metal–Organic Frameworks. Journal of Physical Chemistry C, 2013. 117(43): p. 22250-22259.
34. Kent, C.A., et al., Energy Transfer Dynamics in Metal–Organic Frameworks. Journal of the American Chemical Society, 2010. 132(37): p. 12767-12769.
35. Son, H.J., et al., Light-harvesting and ultrafast energy migration in porphyrin-based metal-organic frameworks. Journal of the American Chemical Society, 2013. 135(2): p. 862-9.
36. Cao, L., et al., Exciton Migration and Amplified Quenching on Two-Dimensional Metal–Organic Layers. Journal of the American Chemical Society, 2017. 139(20): p. 7020-7029.
37. Zhang, Q., et al., Förster Energy Transport in Metal–Organic Frameworks Is Beyond Step-by-Step Hopping. Journal of the American Chemical Society, 2016. 138(16): p. 5308-5315.
38. Son, H.-J., et al., Light-Harvesting and Ultrafast Energy Migration in Porphyrin-Based Metal–Organic Frameworks. Journal of the American Chemical Society, 2013. 135(2): p. 862-869.

39. Liu, J., et al., A novel series of isoreticular metal organic frameworks: realizing metastable structures by liquid phase epitaxy. Scientific Reports, 2012. 2: p. 921.
40. Chernikova, V., O. Shekhah, and M. Eddaoudi, Advanced Fabrication Method for the Preparation of MOF Thin Films: Liquid-Phase Epitaxy Approach Meets Spin Coating Method. ACS Applied Materials & Interfaces, 2016. 8(31): p. 20459-20464.
41. Stavila, V., et al., Kinetics and mechanism of metal–organic framework thin film growth: systematic investigation of HKUST-1 deposition on QCM electrodes. Chemical Science, 2012. 3(5): p. 1531.
42. Summerfield, A., et al., Nucleation and Early Stages of Layer-by-Layer Growth of Metal Organic Frameworks on Surfaces. The Journal of Physical Chemistry C, 2015. 119(41): p. 23544-23551.
43. Zhuang, J.-L., et al., Insight into the Oriented Growth of Surface-Attached Metal–Organic Frameworks: Surface Functionality, Deposition Temperature, and First Layer Order. Journal of the American Chemical Society, 2015. 137(25): p. 8237-8243.
44. Wang, Z., et al., Nanoporous designer solids with huge lattice constant gradients: multiheteroepitaxy of metal-organic frameworks. Nano Lett, 2014. 14(3): p. 1526-9.
45. Liu, J., et al., Photoinduced Charge-Carrier Generation in Epitaxial MOF Thin Films: High Efficiency as a Result of an Indirect Electronic Band Gap? Angew. Chem. Int. Ed., 2015. 54(25): p. 7441-5.
46. Parker, C.A., C.G. Hatchard, and T.A. Joyce, Selective and Mutual Sensitization of Delayed Fluorescence. Nature, 1965. 205(4978): p. 1282-1284.
47. Borisov, S.M., C. Larndorfer, and I. Klimant, Triplet-Triplet Annihilation-Based Anti-Stokes Oxygen Sensing Materials with a Very Broad Dynamic Range. Advanced Functional Materials, 2012. 22(20): p. 4360-4368.
48. Kondakov, D.Y., et al., Triplet annihilation exceeding spin statistical limit in highly efficient fluorescent organic light-emitting diodes. Journal of Applied Physics, 2009. 106(12): p. 124510.
49. Askes, S.H.C., et al., Triplet–triplet annihilation upconversion followed by FRET for the red light activation of a photodissociative ruthenium complex in liposomes †Electronic supplementary information (ESI) available. See DOI: 10.1039/c5cp04352b. Physical Chemistry Chemical Physics, 2015. 17(41): p. 27380-27390.

50. Wohnhaas, C., et al., Triplet–Triplet Annihilation Upconversion Based Nanocapsules for Bioimaging Under Excitation by Red and Deep-Red Light. Macromolecular Bioscience, 2013. 13(10): p. 1422-1430.
51. Monguzzi, A., et al., Photocatalytic Water-Splitting Enhancement by Sub-Bandgap Photon Harvesting. ACS Applied Materials & Interfaces, 2017.
52. Wang, J., et al., Investigation on photocatalytic degradation of ethyl violet dyestuff using visible light in the presence of ordinary rutile TiO_2 catalyst doped with upconversion luminescence agent. Water Research, 2006. 40(11): p. 2143-2150.
53. Lissau, J.S., J.M. Gardner, and A. Morandeira, Photon Upconversion on Dye-Sensitized Nanostructured ZrO_2 Films. Journal of Physical Chemistry C, 2011. 115(46): p. 23226-23232.
54. Lissau, J.S., et al., Anchoring Energy Acceptors to Nanostructured ZrO_2 Enhances Photon Upconversion by Sensitized Triplet-Triplet Annihilation Under Simulated Solar Flux. Journal of Physical Chemistry C, 2013. 117(28): p. 14493-14501.
55. Lissau, J.S., et al., Photon Upconversion from Chemically Bound Triplet Sensitizers and Emitters on Mesoporous ZrO_2: Implications for Solar Energy Conversion. Journal of Physical Chemistry C, 2015. 119(46): p. 25792-25806.
56. Hill, S.P., et al., Photon Upconversion and Photocurrent Generation via Self-Assembly at Organic-Inorganic Interfaces. Journal of Physical Chemistry Letters, 2015. 6(22): p. 4510-4517.
57. Dilbeck, T., S.P. Hill, and K. Hanson, Harnessing molecular photon upconversion at sub-solar irradiance using dual sensitized self-assembled trilayers. Journal of Materials Chemistry A, 2017. 5(23): p. 11652-11660.
58. Cheng, Y.Y., et al., On the efficiency limit of triplet-triplet annihilation for photochemical upconversion. Physical Chemistry Chemical Physics, 2010. 12(1): p. 66-71.
59. Deng, F., J. Blumhoff, and F.N. Castellano, Annihilation Limit of a Visible-to-UV Photon Upconversion Composition Ascertained from Transient Absorption Kinetics. Journal of Physical Chemistry A, 2013. 117(21): p. 4412-4419.
60. Islangulov, R.R., D.V. Kozlov, and F.N. Castellano, Low power upconversion using MLCT sensitizers. Chemical Communications, 2005(30): p. 3776-3778.

61. Baluschev, S., et al., Up-conversion fluorescence: noncoherent excitation by sunlight. Phys Rev Lett, 2006. 97(14): p. 143903.
62. Singh-Rachford, T.N. and F.N. Castellano, Photon upconversion based on sensitized triplet–triplet annihilation. Coordination Chemistry Reviews, 2010. 254(21–22): p. 2560-2573.
63. Monguzzi, A., et al., Low power, non-coherent sensitized photon up-conversion: modelling and perspectives. Phys Chem Chem Phys, 2012. 14(13): p. 4322-32.
64. Monguzzi, A., et al., Upconversion-induced fluorescence in multicomponent systems: Steady-state excitation power threshold. Physical Review B, 2008. 78(19).
65. Singh-Rachford, T.N. and F.N. Castellano, Low Power Visible-to-UV Upconversion. Journal of Physical Chemistry A, 2009. 113(20): p. 5912-5917.
66. Singh-Rachford, T.N., R.R. Islangulov, and F.N. Castellano, Photochemical Upconversion Approach to Broad-Band Visible Light Generation. The Journal of Physical Chemistry A, 2008, 112(17): p. 3906-3910.
67. Murakami, Y., Photochemical photon upconverters with ionic liquids. Chemical Physics Letters, 2011. 516(1-3): p. 56-61.
68. Murakami, Y., H. Kikuchi, and A. Kawai, Kinetics of Photon Upconversion in Ionic Liquids: Time-Resolved Analysis of Delayed Fluorescence. Journal of Physical Chemistry B, 2013. 117(17): p. 5180-5187.
69. Murakami, Y., H. Kikuchi, and A. Kawai, Kinetics of Photon Upconversion in Ionic Liquids: Energy Transfer between Sensitizer and Emitter Molecules. Journal of Physical Chemistry B, 2013. 117(8): p. 2487-2494.
70. Murakami, Y., T. Ito, and A. Kawai, Ionic Liquid Dependence of Triplet-Sensitized Photon Upconversion. Journal of Physical Chemistry B, 2014. 118(49): p. 14442-14451.
71. Hisamitsu, S., N. Yanai, and N. Kimizuka, Photon-Upconverting Ionic Liquids: Effective Triplet Energy Migration in Contiguous Ionic Chromophore Arrays. Angewandte Chemie-International Edition, 2015. 54(39): p. 11550-11554.
72. Turshatov, A., et al., Synergetic Effect in Triplet-Triplet Annihilation Upconversion: Highly Efficient Multi-Chromophore Emitter. Chemphyschem, 2012. 13(13): p. 3112-3115.

73. Monguzzi, A., et al., Efficient Broadband Triplet-Triplet Annihilation-Assisted Photon Upconversion at Subsolar Irradiance in Fully Organic Systems. Advanced Functional Materials, 2015. 25(35): p. 5617-5624.
74. Monguzzi, A., et al., Broadband Up-Conversion at Subsolar Irradiance: Triplet Triplet Annihilation Boosted by Fluorescent Semiconductor Nanocrystals. Nano Letters, 2014. 14(11): p. 6644-6650.
75. Keivanidis, P.E., et al., Up-conversion photoluminescence in polyfluorene doped with metal (II)-octaethyl porphyrins. Advanced Materials, 2003. 15(24): p. 2095-+.
76. Baluschev, S., et al., Enhanced operational stability of the up-conversion fluorescence in films of palladium-porphyrin end-capped poly(pentaphenylene). Chemphyschem, 2005. 6(7): p. 1250-1253.
77. Baluschev, S., et al., Upconversion photoluminescence in poly(ladder-type-pentaphenylene) doped with metal (II)-octaethyl porphyrins. Applied Physics Letters, 2005. 86(6): p. 061904.
78. Baluschev, S., et al., Metal-Enhanced Up-Conversion Fluorescence: Effective Triplet–Triplet Annihilation near Silver Surface. Nano Letters, 2005. 5(12): p. 2482-2484.
79. Jankus, V., et al., Energy Upconversion via Triplet Fusion in Super Yellow PPV Films Doped with Palladium Tetraphenyltetrabenzo-porphyrin: a Comprehensive Investigation of Exciton Dynamics. Advanced Functional Materials, 2013. 23(3): p. 384-393.
80. Lee, S.H., et al., Light upconversion by triplet-triplet annihilation in diphenylanthracene-based copolymers. Polymer Chemistry, 2014. 5(24): p. 6898-6904.
81. Ogawa, T., et al., Highly Efficient Photon Upconversion in Self-Assembled Light-Harvesting Molecular Systems. Scientific Reports, 2015. 5.
82. Avakian, P. and R.E. Merrifield, Triplet Excitons in Anthracene Crystals—A Review. Molecular Crystals, 1968. 5(1): p. 37-77.
83. Duan, P., et al., Photon Upconversion in Supramolecular Gel Matrixes: Spontaneous Accumulation of Light-Harvesting Donor–Acceptor Arrays in Nanofibers and Acquired Air Stability. Journal of the American Chemical Society, 2015. 137(5): p. 1887-1894.
84. Zimmermann, J., et al., Improving triplet-triplet-annihilation based upconversion systems by tuning their topological structure. Journal of Chemical Physics, 2014. 141(18).

85. de Mello, J.C., H.F. Wittmann, and R.H. Friend, An improved experimental determination of external photoluminescence quantum efficiency. Advanced Materials, 1997. 9(3): p. 230-232.
86. Wilson, L.R. and B.S. Richards, Measurement method for photoluminescent quantum yields of fluorescent organic dyes in polymethyl methacrylate for luminescent solar concentrators. Applied Optics, 2009. 48(2): p. 212-220.
87. HAMAMATSU PHOTONICS, K.K., Guide to Streak Cameras. 2008, https://www.hamamatsu.com/resources/pdf/sys/SHSS0006E_STREAK.pdf: HAMAMATSU.
88. Giordmaine, J.A. and R.C. Miller, Tunable Coherent Parametric Oscillation in LiNb${\mathrm{O}}_{3}$ at Optical Frequencies. Physical Review Letters, 1965. 14(24): p. 973-976.
89. Kim, D., Time resolved measurements on Triplet-triplet annihilation based upconversion, in Department of Electrical Engineering and Information Technology 2017, Karlsruhe Institute of Technology. p. 56.
90. Hamamatsu. Photocathode technology. 2018 [cited 2018.
91. Holzwarth, U. and N. Gibson, The Scherrer equation versus the 'Debye-Scherrer equation&#39. Nature Nanotechnology, 2011. 6: p. 534.
92. Oldenburg, M., et al., Enhancing the photoluminescence of surface anchored metal-organic frameworks: mixed linkers and efficient acceptors. Physical Chemistry Chemical Physics, 2018.
93. Oldenburg, M., et al., Photon Upconversion at Crystalline Organic-Organic Heterojunctions. Advanced materials, 2016. 28(38): p. 8477-8482.
94. Arslan, H.K., et al., High-Throughput Fabrication of Uniform and Homogenous MOF Coatings. Advanced Functional Materials, 2011. 21(22): p. 4228-4231.
95. Falcaro, P., et al., Centimetre-scale micropore alignment in oriented polycrystalline metal-organic framework films via heteroepitaxial growth. Nature Materials, 2016. 16: p. 342–348.
96. Dhakshinamoorthy, A., A.M. Asiri, and H. Garcia, Mixed-metal or mixed-linker metal organic frameworks as heterogeneous catalysts. Catal. Sci. Technol., 2016. 6(14): p. 5238-5261.
97. de Juan, A. and R. Tauler, Multivariate Curve Resolution (MCR) from 2000: Progress in Concepts and Applications. Critical Reviews in Analytical Chemistry, 2006. 36(3-4): p. 163-176.

98. Kanoo, P., et al., Coordination driven axial chirality in a microporous solid assembled from an achiral linker via in situ C-N coupling. Chemical Communications, 2011. 47(39): p. 11038-11040.
99. Correa, N.M. and N.E. Levinger, What Can You Learn from a Molecular Probe? New Insights on the Behavior of C343 in Homogeneous Solutions and AOT Reverse Micelles. The Journal of Physical Chemistry B, 2006. 110(26): p. 13050-13061.
100. Reynolds G. A., D.K.H., New coumarin dyes with rigidized structure for flashlamp-pumped dye lasers. Optics Communications, 1975. 13: p. 222-225.
101. Zhang, Q., et al., Forster Energy Transport in Metal-Organic Frameworks Is Beyond Step-by-Step Hopping. J. Am. Chem. Soc., 2016. 138(16): p. 5308-15.
102. Zaikowski, L., et al., Charge Transfer Fluorescence and 34 nm Exciton Diffusion Length in Polymers with Electron Acceptor End Traps. The Journal of Physical Chemistry B, 2015. 119(24): p. 7231-7241.
103. Mikhnenko, O.V., P.W.M. Blom, and T.-Q. Nguyen, Exciton diffusion in organic semiconductors. Energy & Environmental Science, 2015. 8(7): p. 1867-1888.
104. Caram, J.R., et al., Room-Temperature Micron-Scale Exciton Migration in a Stabilized Emissive Molecular Aggregate. Nano Lett., 2016. 16(11): p. 6808-6815.
105. Donati, D. and J.O. Williams, Exciton Diffusion Lengths for Pure and Doped Anthracene Single Crystals from Microscopic Measurements. Molecular Crystals and Liquid Crystals, 1978. 44(1-2): p. 23-32.
106. Kent, C.A., et al., Light Harvesting in Microscale Metal–Organic Frameworks by Energy Migration and Interfacial Electron Transfer Quenching. Journal of the American Chemical Society, 2011. 133(33): p. 12940-12943.
107. Tamai, Y., et al., Exciton Diffusion in Conjugated Polymers: From Fundamental Understanding to Improvement in Photovoltaic Conversion Efficiency. Journal of Physical Chemistry Letters, 2015. 6(17): p. 3417-28.
108. Oldenburg, M., et al., Photon Upconversion at Crystalline Organic-Organic Heterojunctions. Advanced materials, 2016. 28(38): p. 8477-8482.
109. de Miguel, G., et al., Tunable soret-band splitting of an amphiphilic porphyrin by surface pressure. Chemphyschem, 2008. 9(11): p. 1511-3.

Schriften des Instituts für Mikrostrukturtechnik
am Karlsruher Institut für Technologie (KIT)

ISSN 1869-5183

Herausgeber: Institut für Mikrostrukturtechnik

Band 1 **Georg Obermaier**
Research-to-Business Beziehungen: Technologietransfer durch Kommunikation von Werten (Barrieren, Erfolgsfaktoren und Strategien). 2009
ISBN 978-3-86644-448-5

Band 2 **Thomas Grund**
Entwicklung von Kunststoff-Mikroventilen im Batch-Verfahren. 2010
ISBN 978-3-86644-496-6

Band 3 **Sven Schüle**
Modular adaptive mikrooptische Systeme in Kombination mit Mikroaktoren. 2010
ISBN 978-3-86644-529-1

Band 4 **Markus Simon**
Röntgenlinsen mit großer Apertur. 2010
ISBN 978-3-86644-530-7

Band 5 **K. Phillip Schierjott**
Miniaturisierte Kapillarelektrophorese zur kontinuierlichen Überwachung von Kationen und Anionen in Prozessströmen. 2010
ISBN 978-3-86644-523-9

Band 6 **Stephanie Kißling**
Chemische und elektrochemische Methoden zur Oberflächenbearbeitung von galvanogeformten Nickel-Mikrostrukturen. 2010
ISBN 978-3-86644-548-2

Die Bände sind unter www.ksp.kit.edu als PDF frei verfügbar oder als Druckausgabe zu bestellen.

Band 7 Friederike J. Gruhl
Oberflächenmodifikation von Surface Acoustic Wave (SAW) Biosensoren für biomedizinische Anwendungen. 2010
ISBN 978-3-86644-543-7

Band 8 Laura Zimmermann
Dreidimensional nanostrukturierte und superhydrophobe mikrofluidische Systeme zur Tröpfchengenerierung und -handhabung. 2011
ISBN 978-3-86644-634-2

Band 9 Martina Reinhardt
Funktionalisierte, polymere Mikrostrukturen für die dreidimensionale Zellkultur. 2011
ISBN 978-3-86644-616-8

Band 10 Mauno Schelb
Integrierte Sensoren mit photonischen Kristallen auf Polymerbasis. 2012
ISBN 978-3-86644-813-1

Band 11 Daniel Auernhammer
Integrierte Lagesensorik für ein adaptives mikrooptisches Ablenksystem. 2012
ISBN 978-3-86644-829-2

Band 12 Nils Z. Danckwardt
Pumpfreier Magnetpartikeltransport in einem Mikroreaktionssystem: Konzeption, Simulation und Machbarkeitsnachweis. 2012
ISBN 978-3-86644-846-9

Band 13 Alexander Kolew
Heißprägen von Verbundfolien für mikrofluidische Anwendungen. 2012
ISBN 978-3-86644-888-9

Band 14 Marko Brammer
Modulare Optoelektronische Mikrofluidische Backplane. 2012
ISBN 978-3-86644-920-6

Band 15 Christiane Neumann
Entwicklung einer Plattform zur individuellen Ansteuerung von Mikroventilen und Aktoren auf der Grundlage eines Phasenüberganges zum Einsatz in der Mikrofluidik. 2013
ISBN 978-3-86644-975-6

Band 16 Julian Hartbaum
Magnetisches Nanoaktorsystem. 2013
ISBN 978-3-86644-981-7

Band 17 Johannes Kenntner
Herstellung von Gitterstrukturen mit Aspektverhältnis 100 für die Phasenkontrastbildgebung in einem Talbot-Interferometer. 2013
ISBN 978-3-7315-0016-2

Band 18 Kristina Kreppenhofer
Modular Biomicrofluidics - Mikrofluidikchips im Baukastensystem für Anwendungen aus der Zellbiologie. 2013
ISBN 978-3-7315-0036-0

Band 19 Ansgar Waldbaur
Entwicklung eines maskenlosen Fotolithographiesystems zum Einsatz im Rapid Prototyping in der Mikrofluidik und zur gezielten Oberflächenfunktionalisierung. 2013
ISBN 978-3-7315-0119-0

Band 20 Christof Megnin
Formgedächtnis-Mikroventile für eine fluidische Plattform. 2013
ISBN 978-3-7315-0121-3

Band 21 Srinivasa Reddy Yeduru
Development of Microactuators Based on the Magnetic Shape Memory Effect. 2013
ISBN 978-3-7315-0125 1

Band 22 Michael Röhrig
Fabrication and Analysis of Bio-Inspired Smart Surfaces. 2014
ISBN 978-3-7315-0163-3

Band 23 Taleieh Rajabi
Entwicklung eines mikrofluidischen Zweikammer-Chipsystems mit integrierter Sensorik für die Anwendung in der Tumorforschung. 2014
ISBN 978-3-7315-0220-3

Band 24 Frieder Märkle
Laserbasierte Verfahren zur Herstellung hochdichter Peptidarrays. 2014
ISBN 978-3-7315-0222-7

Band 25 Tobias Meier
Magnetoresistive and Thermoresistive Scanning Probe Microscopy with Applications in Micro- and Nanotechnology. 2014
ISBN 978-3-7315-0253-1

Band 26 Felix Marschall
Entwicklung eines Röntgenmikroskops für Photonenenergien von 15 keV bis 30 keV. 2014
ISBN 978-3-7315-0263-0

Band 27 Leonardo Pires Carneiro
Development of an Electrochemical Biosensor Platform and a Suitable Low-Impedance Surface Modification Strategy. 2014
ISBN 978-3-7315-0272-2

Band 28 Sebastian Mathias Schillo
Prozessentwicklung für die Automatisierung der Herstellung und Anwendung von hochdichten Peptidmicroarrays. 2014
ISBN 978-3-7315-0274-6

Band 29 Nicole E. Steidle
Micro- and Nanostructured Microfluidic Devices for Localized Protein Immobilization and Other Biomedical Applications. 2014
ISBN 978-3-7315-0297-5

Band 30 Jochen Heneka
Prozessentwicklung eines industrietauglichen Verfahrens zur Fertigung von vereinzelten LIGA-Mikrobauteilen. 2015
ISBN 978-3-7315-0326-2

Band 31 Seoung-Eun Kim
Konzeption und prototypische Fertigung einer nicht-invasiven mikrofluidischen Plattform für die Elektrophysiologie (NIMEP) zur Zellenanlyse. 2015
ISBN 978-3-7315-0378-1

Band 32 Elisabeth Wilhelm
Entwicklung eines mikrofluidischen Brailledisplays. 2015
ISBN 978-3-7315-0385-9

Band 33 Viktor Pinneker
Entwicklung miniaturisierter Aktorsysteme basierend auf magnetischen Formgedächtnislegierungen. 2018
ISBN 978-3-7315-0500-6

Band 34 Ali Caglar Özen
Novel MRI Technologies for Structural and Functional Imaging of Tissues with Ultra-short T_2 Values. 2017
ISBN 978-3-7315-0657-7

Band 35 Anne Bäcker
Veränderliche 3D Zellgerüstträger auf Cryogelbasis zur Kultivierung von Prostatakarzinomzellen. 2017
ISBN 978-3-7315-0676-8

Band 36 Frieder Johannes Koch
X-ray optics made by X-ray lithography: Process optimization and quality control. 2017
ISBN 978-3-7315-0679-9

Band 37 Tobias Jörg Schröter
Vergrößerung des Sehfeldes der Röntgen-Phasenkontrast-Bildgebung für die klinische Anwendung. 2017
ISBN 978-3-7315-0731-4

Band 38 Felix Vüllers
Bioinspired Superhydrophobic Nano- and Microstructured Surfaces for Drag Reduction and Optoelectronics. 2018
ISBN 978-3-7315-0816-8

Band 39 Frederik Kotz
Entwicklung neuer Materialien für die additive Fertigung und das Rapid Prototyping von Glas und Polymethylmethacrylat. 2018
ISBN 978-3-7315-0835-9

Band 40 Michael Oldenburg
Photon upconversion heterostructures made from surface-anchored metal-organic frameworks. 2019
ISBN 978-3-7315-0863-2

Die Bände sind unter www.ksp.kit.edu als PDF frei verfügbar oder als Druckausgabe zu bestellen.